THE ROAD TO NOW

THE ROAD TO NOW

Taking stock of evolution and our place in the world

MELVIN BOLTON

ALLEN&UNWIN

First published in 2001

Allen & Unwin
83 Alexander Street
Crows Nest NSW 2065
Australia
Phone: (61 2) 8425 0100
Fax: (61 2) 9906 2218
Email: info@allenandunwin.com
Web: http://www.allenandunwin.com

National Library of Australia
Cataloguing-in-Publication entry:

Bolton, Melvin.
The road to now: taking stock of evolution and our place in the world.

  Bibliography.
  Includes index.

  ISBN 1 86508 490 5

  1. Evolution (Biology). I. Title. II. Title: taking stock of evolution and our place in the world.

576.8

Set in 11/13.5 pt Janson by Midland Typesetters, Maryborough
Printed by South Wind Production Ltd, Singapore

CONTENTS

Preface ix

1. KNOWING AND BELIEVING 1
   Big Old Questions 1
   Early Kinds of Answers 3
   Greeks Were Different 4
   Biology After Aristotle 7
   Science Breaks Through 9
   Being Scientific 12
   Numbers and Theories 14

2. WHAT IS THIS THING CALLED LIFE? 18
   Vital Sparks and Candle Flames 18
   Life's Ingredients 21
   A Necessary Bit About Chemistry 22
   Nucleic Acids and the Language of Heredity 25
   Chromosomes 28
   Genes 31

3. BEGINNINGS 34
   Cosmic Productions 34
   Life's Origins: Early Experiments 36
   More Possibilities 38
   A Mainstream Position 41

4. CHEMICALS TO CREATURES 44
   Early Cells 44
   Early Sex 45
   Bacteria and a Point of No Return 46
   All That Are Not Bacteria 47
   Those Organelles 50
   Multicells: The End of Immortality 52

5. DARWIN'S BEST SELLER   56
   The Doctor's Son   57
   Rocks and a Wedding   59
   The Man and His Book   60
   What Darwin Didn't Know   64
   The New Synthesis   68
   Deceptive Simplicity   69

6. DARWINIANS WITH COMPUTERS   72
   Evolution of Species   73
   Eyeballs and Sickle Cells   74
   Since the Modern Synthesis   77
   Cautionary Tales and Confusions   80
   Being Good: A Real Problem   84
   Sex: Another Problem   87

7. SHAPING OUR FAMILY TREE   89
   Milk, Hair and Jawbones   91
   Enter the Primates   92
   Up on Two Legs   94
   Missing Links and Broken Chains   96
   Homo   98
   A Confusion of Humans   99
   One Species Left   102
   Neanderthals and Us   103

8. WERE PEOPLE EVER WILDLIFE?   105
   Forces of Change   106
   Adapted Man   107
   Cerebral Man   109
   Handy Man   111
   Social Man   112
   Family Man   116
   Symbols and Language   119
   Human Wildlife?   120

9. ALTERING THE LANDSCAPE   122
   Ecological Man   124
   People Shaping Up   126
   Working Through the Stone Age   128

The Farming Life 130
City-states to Nation-states 132
Urban Man 134

10. DARWIN, GOD AND SOCIETY 137
    Souls and Savages 138
    Darwin's Reception 141
    Darwin to the Right 143
    Darwin to the Left 146
    Evolution Everywhere 149

11. EXPLOSIVE TIMES: HUMAN ECOLOGY IN THE
    TWENTIETH CENTURY 152
    Ages and Numbers 153
    Feeding Ourselves 154
    Simplifying the Earth 158
    Lifestyles and Energy 159
    Pollution, Degradation and Confusion 160
    Who Cares, and Why? 165

12. RECIPES AND OUTCOMES 168
    Reading the Cookery Book 168
    Genes at Work 170
    Nature and Nurture 171
    Heritability and the Value of Twins 173
    Twins, Frauds and Strong Feelings 175
    Studies of Twins After Burt 177
    Intelligence 179

13. GENES AND BEHAVIOUR 185
    Temperament and Personality 185
    Once Upon a Time ... 188
    Dogs, Cats, Pigeons and Utopia 190
    Running on Instinct 193
    Instincts and Opportunities 196
    Modules of the Mind 198
    Time Scales 200

14. THE SOCIAL DIMENSION 202
    The Failure to Communicate 203

Time Scales and Consciousness   204
Themes and Variations   206
The Big Taboo   209
Blood is Thicker ...   210
Doing Unto Others   211
Social Modules   213
Outwitting the Beast: Emotions and Morals   215

15. OLD BRAINS, NEW BRAINS   218
Layout of the Brain   220
Mainly Human   222
Learning and Memory   224
Seeing is Believing   228
Left Brains and Right Brains   229
Male and Female   231
The Ultimate Enigma   234

16. THE FUTURE IS COMING   237
Population Changes in the Modern World   238
Lifting and Shifting the Pressures   242
Environmental Change and Population Density   243
People in Control   245
Gods in Control   247
A Final Thought   251

Notes   252
Select Bibliography   283
Index   285

PREFACE

I once saw a list of household items that, before about 1950, we managed quite well to live without. They were things like plastic bags, ballpoint pens, credit cards, frozen foods and television. It was a reminder of how much our domestic lives had changed over a few decades. In contrast, for well over a million years the cutting edge of technology was a single type of stone axe.

Compared with our Stone Age conservatism, the technological blast-off of the past half century is beyond all superlatives. But modern science and technology have not been applied only to the things we use—we also know a lot more about the world and about ourselves than we did in the 1950s. Today's bookshops are stacked with volumes written by distinguished scientists and aimed at the general reader. There has been a corresponding increase in the number of journals and magazines, and of course there is now television and the Internet.

So those of us who live in western societies ought to be much better informed and more scientifically minded than we were 50 years ago, as well as being better equipped and provisioned. But are we? Could it not be that a great many people are now experiencing an information overload? How many of us can find the time to follow the arguments of dozens of specialists in their own fields?

It is inevitable that what were once well-integrated studies in the sciences and humanities will continue to divide into ever-narrower specialisations as knowledge grows. But not all academics are happy about this. The downside of the trend was the theme of a millennium essay in the prestigious science journal *Nature*.[1] The author lamented the fact that many researchers know nothing about the research being done in areas outside their immediate professional concern. And if the specialists are losing sight of the broader picture, it can be of no help to the rest of us.

Scientists, of course, are subject to the same passions, persuasions and aversions that affect other people, so they have different ideas about how their work should be presented to the general public. And regardless of what the scientists may think and say, they cannot prevent their reports from being distorted or sensationalised as news gets relayed through the more popular media, especially when the research findings are complex and the issues involved emotive. As a consequence, it is easy to believe that today's readers are finding it hard to know where mainstream science stands, especially in that broad interface between biology and the social sciences. It was in that belief that I decided to add yet another book to the long list that comes under the heading of popular science. I hope it will be a book that clarifies and does not add to the confusion.

Throughout the book we shall generally be following the path of established knowledge or mainstream scientific thinking, though we shall not be afraid of meeting dissenters and mavericks when they have something interesting to say. We shall encounter disagreement and controversy and try to see why it exists rather than following one side of the argument as if there were no dispute. We shall also enter, but knowingly so, the realm of pure speculation. Sources of information and suggestions for further reading will be listed for each chapter, but I don't want to bombard readers with hundreds of references to scientific papers published in obscure journals, so as far as possible I shall refer to publications that the general reader has a reasonable chance of finding.

I believe it is important, in a book of this sort, to find the right scale: to try to include enough detail to fill the field of view without any clutter, but also without troubling omissions. We choose our scale according to our purpose. One standard textbook on the structure of the gene runs to 1200 pages, whereas the Chambers Dictionary offers a meaningful definition of a gene in 23 words. I shall be much closer to the dictionary than to the textbook, so I must not lose sight of my purpose. The intention is to put the history of our species in a biological context through the medium of plain language. By the end I hope that the non-biologist readers who have stayed with me will better appreciate what biology can tell us about our place in nature and will feel able to scan the bookshelves with more confidence than they had before. I hope they will feel able to consider critically why one book claims that evolution is a theory in crisis while others affirm that neo-Darwinism is more solid than the theory of gravity. Or why one book tells us how little is determined by our genes while others suggest that we can be regarded as mere robots for carrying out their program. I should say here that I am using the word biology in the widest possible sense, but we will sometimes need to roam even more widely than that. My purpose is to interpret a story that is both fascinating and disturbing, and it should not be limited by the constraints of academic demarcation lines. I shall do my best to get the scale right, but others must judge how far I succeed.

Except for a few insertions and amendments, I wrote the previous paragraphs more than two years ago, and it remains only for me to thank those who have helped me to get the writing finished. Early in the project I received some useful pointers from three anonymous referees who had to base their judgements on a couple of sample chapters and a brief outline. I tried to follow their advice and am grateful to them, but I make the usual disclaimer on their behalf and for all others who helped me. Any errors are mine.

At intervals I had to appeal to researchers and others for information, and I am pleased to be able to thank the following for their prompt responses: Terry Schwaner, Kwang Jeon, Mark Collard, Colin Groves, Roger Highfield, Eleanor Maguire, Thomas Bouchard, Jeremy Speck, David Buss, Glenn Weisfeld and Jenny Saunders.

The draft was improved by the perceptions and input of those readers who made the effort for me when they had more pressing things to be doing. Thank you Margaret Curley, Julie Maher, Robert Clark, Hilary Bolton and David Rogers.

Finally, I thank my wife, Jean, for her unfailing help and support in this, as in all my endeavours.

MB
July 2000

CHAPTER 1

KNOWING AND BELIEVING

A SKING QUESTIONS IS EASY. Asking sensible questions, to get the sort of answers you had in mind, needs a bit more thought. Just because a question can be framed doesn't mean that there has to be a sensible answer. After all, one could enquire, with impeccable syntax, about curriculum development in the aeronautical academy for pigs.

About ten years ago the editors of *Life* magazine asked 'What is the meaning of life?', and invited several hundred of 'the world's less ignorant men and women to come up with a few answers'. The 300 contributions later published in book form range from the hauntingly eloquent to the comical.[1] That was what the editors wanted, of course. The question is completely open-ended and it begs the more fundamental queries: is there reason to think that life has meaning in human terms, and is there any one meaning of special significance for us? Even if we break the question down in this way it can still be approached from a philosophical, scientific, religious or flippant point of view. Well, these days it can. It hasn't always been so.

BIG OLD QUESTIONS

There is ample evidence that people were concerned about life's deeper mysteries long before science, philosophy or theology existed in any modern sense of the words. Nobody can know precisely what sort of questions most troubled the people of the Old Stone Age, but death must surely have been an awful mystery from the time people were able to reflect on it. Edward B. Tylor, one of the founders of anthropology, pointed out in 1881 that

1

death was probably not always seen as an all-or-nothing phenomenon:

> A person who a few minutes ago was walking and talking, with all his senses active, goes off motionless and unconscious in a deep sleep, to awake after a while with renewed vigour. In other conditions the life ceases more entirely, when one is stunned or falls into a swoon or trance, where the beating of the heart and breathing seem to stop, and the body, lying deathly pale and insensible, cannot be awakened; this may last for minutes or hours, or even days, and yet after all the patient revives. Barbarians are apt to say that such a one died for a while, but his soul came back again. They have great difficulty in distinguishing real death from such trances. They will talk to a corpse, try to rouse it and even feed it, and only when it becomes noisome and must be got rid of from among the living, they are at last certain that the life has gone never to return. What, then, is this soul or life which thus goes and comes in sleep, trance, and death? To the rude philosopher, the question seems to be answered by the very evidence of his senses. When the sleeper awakens from a dream, he believes he has really somehow been away, or that other people have come to him. As it is well known by experience that men's bodies do not go on these excursions, the natural explanation is that every man's living self or soul is his phantom or image, which can go out of his body and see and be seen itself in dreams.[2]

People who survived until recently with only Stone Age technology can teach us a lot about ancient ways of thinking. The Australian Aborigines traditionally see themselves as such an integral part of their environment that in their mythology they are inseparable from other creatures and from the inanimate features of their traditional lands—such as the rivers, hills and rocks. The coming into being and existing and departing of all these elements are woven into the common fabric of substance and spirit.[3] This way of thinking is called animism, from the Latin root *anima*, meaning 'soul' or 'life'. In this view, even non-living things may have a spirit or be inhabited by the spirits of ancestors. And why not? They also occur in dreams. Animism as a world view takes many forms, but was encountered by missionaries in most of the preliterate societies of Africa, Asia, Oceania and the Americas. Whether it should be regarded as the original religion, as Tylor believed, is open to dispute and fortunately isn't a question we need to struggle with here. The point to be taken is that although we humans may always have been concerned with similar big issues, our way of questioning, and the actual questions we ask, are very much determined by the background knowledge and beliefs of each society.

EARLY KINDS OF ANSWERS

For animists there was no distinction between the natural and the supernatural. This is still true for the few remaining people who have not been much influenced by modern cultures, and it is not difficult to understand. If the only world view you have ever heard is that you and all your people came from the belly of a mountain and that the spirit of that mountain will determine your destinies, influence your successes and failures, your loves and hopes and those of your offspring, then those are your facts of life. Knowing them and believing them are the same thing. Notions of the supernatural can have no meaning because nature, as you know it, has all the wondrous powers imaginable. The fact that an outsider can violate the mountain, break all your rules and get away with it only proves what your people, the children of the mountain, already knew: that the outsider is not of your kin. For him, there are different facts of life.

The oneness of the natural and the supernatural, or divine, was not questioned by the first great civilisations of about 5000 years ago, but it was a world view which by then had grown in complexity. The Sumerians of Mesopotamia had several thousand deities which accounted for everything that needed an explanation—including human creation, fertility, crop failure, disease and death. The people of ancient Egypt, who founded a much more durable civilisation, also had a bewildering collection of gods who could be called on for every circumstance or held responsible for any eventuality. Some of the gods have been traced back to prehistoric times. Many of the gods and their complicated relationships were intimately connected with the mammals, birds and reptiles known to the ancient Egyptians.

The live animals themselves were not necessarily regarded as gods, but they were seen as the earthly representations of the deities they physically resembled. Aman, for example, the king of the gods, is shown as a man with a ram's head or a ram with a crown, while Bast or Bastet, the goddess of music and dance, is a cat or a cat-headed woman.[4] Depending on the importance of the deity involved, the corresponding living animals were held sacred, such that severe penalties—including death—were imposed on anybody who deliberately killed one. Today, in Hindu symbolism, the cow has sacred significance as a mother figure, but, to the relief of one motorist I know, the death penalty is not imposed on drivers who hit one on the road. The unhappy offender was only locked up for a few days.

In summary, from the evidence of early carvings and cave art, from history and from people who still follow the hunter–gatherer traditions, scholars have put together a general under-standing of the way people saw themselves and their societies in relation to the worlds they knew in the distant past. There is enormous variation in the detail but the basis for all truth almost everywhere seems to have been the authority of tradition. All human affairs, the routines and the life crises, and all relationships with the external world were to be explained and understood in the same terms—the terms dictated by higher powers.

GREEKS WERE DIFFERENT

From ancient Greece came an entirely new way of looking at nature. One recent author, John Moore, refers to this sudden appearance of naturalistic thought as 'one of the most astonishing events in intellectual history'.[5] Following the collapse of the Bronze Age civilisations in Greece, about 3000 years ago, Athens became the cultural centre. The Greek colonial expansions were still to come but the port of Miletus on the coast of Ionia (now Turkey) had already been settled by Greeks. During the sixth century one of these Milesians, a philosopher called Thales (c. 625–547 BC), is credited with being the first person known to have systematically searched for explanations of nature on the basis of his observations and reasoning instead of attributing everything to gods and spirits. Thales and two of his students were particularly concerned with finding a substance that was common to all matter, the universal stuff of nature. Thales supposed it must be water but one of his students (Anaximenes) thought it more likely to be air.[6] Their conclusions are not important. The point is that in their approach we can recognise the origin of rational thought. Not everybody would nominate Thales of Miletus as the man who set the West on its scientific course, but at the very least he and his students bridged the gap between mythology and a new attitude of critical enquiry.

A century or so later Hippocrates (c. 460–377 BC) and his contemporaries established a medical tradition based on organic explanations in which demons played no part. According to the Hippocratic school a healthy body was one with the correct equilibrium in the proportions of blood, phlegm, yellow bile and black bile. An excess of one of them made the patient, respectively, sanguine, phlegmatic, choleric or melancholic. It was important to record the symptoms and course of an illness with a view to determining which of the four bodily humours was out of balance. Recovery and prevention depended on proper rest, exercise and diet to restore the harmony of the body.[7]

Not much is known about Hippocrates as a person, but with Aristotle (384–322 BC) we are on much firmer historical ground, and it was with Aristotle that the science of ancient Greece reached its climax. A doctor's son from northern Greece, Aristotle was sent to the Academy of Plato in Athens as a youth and stayed for 20 years. After leaving the Academy his interest in mathematics and astronomy gave way to an involvement with living things that led to his publishing (among his many other works) several treatises on zoology.[8] In 335 BC he returned to Athens and opened the Lyceum, a centre of learning and research which came to rival Plato's Academy—but specialising in biology and history. To begin with, the Lyceum was an informal affair, with instruction being given in the covered walkways, or *peripatos*, of a temple (which is how we got the word peripatetic for itinerants).

It would be easy, reading about Aristotle's output of descriptive biology, to miss the important point of it all. Simply compiling facts is not science; if it were, then a phone directory would be a scientific achievement. Aristotle's *Historia animalium* covers the structure and habits of an estimated 500 different species of animals, but it is Aristotle's attempt at understanding, based on his systematic observations, that makes his work truly scientific. He also understood that to get useful answers you had to be careful with your questions.

Aristotle was concerned with several big questions in relation to biology. He was interested in the processes of life and, like generations of people before and after him, he wanted to know what it was that made the difference between being alive and being dead. Vitalism, the idea that there must be some vital principle that imparted life, was the starting point, but what was the vital principle? Plato had reasoned that body and soul have a dual existence and that the soul existed before the birth of a body and after its death, but he had arrived at that *dualist* view only with regard to human beings. To Aristotle it was plain that many animals, including humans, have comparable structures and common characteristics. And they all move, eat, breathe, sleep, void waste, grow, breed and die. He recognised the relationship between structure and function (or purpose, as he called it) in those parts of animals that were observable, and he devoted one treatise—*De partibus animalium*—to the significance of the parts of animals. He even arrived at a classification, roughly in line with the vertebrate–invertebrate division, based on the presence or absence of blood. It made sense to him, therefore, to consider all living things according to their common characteristics. Did they also, he asked, have life by virtue of a common vital principle?

In a psychological tract, *De anima*, Aristotle argued that this vital principle or life force was a property of organised bodies and would differ according to the level of organisation of the body. Plants would therefore have the simplest kind of life force and sentient animals would have a higher kind. As to the nature of this life force, he argued that the life force was to the body as the shape, or form, is to the wax; it made no sense to ask whether one could exist without the other. Disembodied souls, therefore, were ruled out. But Aristotle deliberately left a chink in his argument in readily admitting that there was no evidence as to the nature of mind or the power to think. This, it seemed to him, was a widely different kind of soul. It was an admission not lost on the founding fathers of the Christian Church a few centuries later.

Clearly, even with the mind of an Aristotle, your conclusions are only as firm as your evidence, and in posing questions on the essence of life and the nature of mind, or soul, Aristotle had not much more to go on than the next man. At that time, despite the work of the Hippocratic school, even the functions of the major organs were open to speculation. Peering into the body cavities of corpses in those days must have been utterly baffling. What was to be made of it all, this well-packed and colourful assemblage of organs and tubes and slimy membranes? What did it all *do*? In fact, in *De partibus*, Aristotle gives a good account of the function of the kidneys and bladder, having traced the ducts and noted the distribution of blood and urine. It was a level of understanding, as Moore points out, not improved on until the nineteenth century. Aristotle had also traced the passage of food through the stomach and intestines, but he thought digestion was achieved by heat, which came from the heart, with blood as the end product. All he could say about the liver was that it was somehow necessary for making use of the nourishment in the blood.

Most of Aristotle's original writings have been lost and we have to rely on the Lyceum's last director, working more than 260 years after the great man's death, for the editing of the manuscripts that have been preserved. Scholars have been picking over the details for centuries, but nothing will change the fact that Aristotle's influence on Western thought has been immense. In biology he recognised the scope of the subject, identified problems in scientific terms, and was systematic at collecting observations that might throw light on them. More widely, our English vocabulary, with its words from Greek and Latin roots, is sprinkled with particular meanings which Aristotle introduced—words like energy, genus, substance, quality and potential. And yet Aristotle himself probably did not make much of a mark on society. In his time he was an ivory-tower academic, and the scientific tradition that he originated was a long time taking hold.

BIOLOGY AFTER ARISTOTLE

The main flowering of Greek biological science came to an end with the deaths of Aristotle and his students at the Lyceum. The main centres of learning shifted to Alexandria, and biology contracted to the study of the human body—making very slow progress even then. In the time of Galen, a Greek physician who practised in Rome during the second century AD, medicine was still largely based on what the Hippocratic school had established some 600 years earlier—including the doctrine of the humours. Knowledge of anatomy and physiology was still vague. It was known that the brain was the centre of intelligence, and that blood must move in some sort of shuttle system between heart and lungs. Galen correctly concluded that the arteries and veins must be connected in some way even though he couldn't see the capillaries. It was still a mystery how blood vessels carried air from the lungs.[9]

In Rome, Galen was physician to the gladiators, so he must have seen plenty of blood; but the dissection of corpses was not allowed. Galen's own studies of internal anatomy were made on monkeys or pigs, and for the most part his knowledge of human anatomy was still based on the work of the earlier Greek physicians of Alexandria. It was a state of affairs that was to prevail for another 1000 years. Biology effectively came to a halt with Galen. The Romans, for all their advances in government, law, civic works and sanitation, evidently built their empire without furthering their knowledge of basic science. There is no evidence that the Chinese or Indians were any further forward. Their very ancient medical traditions were not based on an understanding of anatomy or physiology, and neither culture permitted the opening of cadavers to satisfy academic curiosity.

By the time the Roman Empire finally fell into anarchy the Catholic Church had become the state religion and was singularly influential. Its version of the truth in all things was largely based on Saint Augustine's interpretation of the scriptures. Saint Augustine of Hippo, who was born about 150 years after Galen's death, believed that everything could be explained by theology, so that scholarship essentially became Bible study. Throughout Christendom this kept the door firmly slammed on biology until the eleventh century.

So it was that the Islamic world took custody of science from the time of Galen until the Middle Ages. It was the Arabs, Persians and others from the Muslim Empire who translated and studied the works of ancient Greece, and it was they who achieved what progress there was during the first millennium. Most scientific advances under Islam were in maths and astronomy but a Persian physician, Ibn Sina, known in the West as Avicenna, produced a classic compilation of Greek and Islamic medicine known as *The Canon of Medicine*. The Islamic scholars introduced many new drugs and methods of preparing medicines, but in human biology they didn't advance the work of Hippocrates, Aristotle and Galen.[10]

When the works of Aristotle were secondarily translated from Arabic into Latin in the Middle Ages, they caused a tumult both in theology and science. Saint Thomas Aquinas (*c.* 1224–1274) synthesised Aristotle's philosophy and existing Christian ideas to come up with an explanation of the soul and human nature that has become Roman Catholic orthodoxy.[11] The early medical schools of Europe, which emerged in Italy and France during the twelfth century, seized on Avicenna's *Canon* and so were introduced to Aristotle with an Islamic overlay. These medical schools, at Salerno, Padua and Montpellier, gradually produced scholars with enough confidence to question the authority of the ancients and to trust their own observations, but the new age of learning in biology had a very slow fuse. For example, although the Renaissance is generally dated from the fourteenth century, it was not until 1543 that Galen's errors in human anatomy were substantially corrected in a new publication by a surgeon at Padua University. The surgeon, a Belgian called Andreas Vesalius, had done his own dissections on cadavers (which he had to steal at night from a public gallows) and had his book illustrated by woodcuts probably made by students of the great artist Titian. It was printed on the then new printing press in Switzerland.[12] By that time Leonardo da Vinci had also gained new insight into anatomy, but his observations were recorded only in his notebooks.

SCIENCE BREAKS THROUGH

It took a lot of nerve to question the establishment in sixteenth century Europe. Church dogma was rigid, and individuals who encouraged the public to question it were liable to be burned to death for heresy. Galen's version of anatomy had been accepted by the Church, as had Ptolemy's version of astronomy from the second century. In the year that Vesalius' anatomy book was printed, the astronomer Copernicus went into print with his claim that the Earth wasn't the centre of the universe but was actually revolving round the sun. Then he died of natural causes and possibly had a lucky escape. Others who openly promoted his ideas didn't fare so well. Giordano Bruno, an Italian philosopher and astronomer, went bravely to the stake in 1600. Galileo was treated with more care but was still forced to recant the Copernican doctrine in 1633. The Church had painted itself into a corner, though, and no approved authority—not even Aristotle—was immune from critical reappraisal in that new age of questioning and discovery that came to be known as the scientific revolution.

One tends to forget that 'science' has not always meant what it means today. In sixteenth century Europe it meant what we now call theology and philosophy. The 'arts'—liberal, learned and mechanical—covered all other knowledge. Languages and maths were liberal arts; medicine and law were learned arts; while the mechanical arts were what we might now call crafts or trade skills, though they included such professional disciplines as navigation, cartography—and surgery.[13] This all makes sense in terms of the prevailing world view of that time. There was no new knowledge, in the sense of things to be discovered that had never been known. Everything was known to God, and in the golden age of the beginning all had been known to man. Indeed, God had created the world *for* people. This was the common sense of the time, and it stood to reason that theology and philosophy were the routes to regaining that lost knowledge and reconciling it with what one perceived through the senses. Truth could be revealed only through God; otherwise how could we ever know it was the truth? Humans had no direct line to the facts of life, they could find them only through God's illumination. Scientific problems, therefore, were theological issues, and any contrary views were not only wrong—they were a threat to the establishment.

The scientific revolution involved a new way of thinking and was something of a creeping change. It must have been difficult for those born in the sixteenth century to regard anything natural as being so lacking in divine plan or purpose that the truth of it could be understood from direct observation. Only gradually did scientific discoveries come to be regarded as something other than religious insight. Indeed, from its roots in the scientific revolution, modern science owes much to this subtle interplay of religion and natural philosophy because the experimenters were often at pains to show how their work supported their theological beliefs. Even Sir Francis Bacon described his efforts to reform natural philosophy as being a preparation for the eternal Sabbath that would follow the Day of Judgement.[14]

Bacon (1561–1626), the barrister, writer and parliamentarian who became England's lord chancellor, made no scientific discoveries but had much to say on the advancement of learning and scientific method. He warned against the traditional process and advocated a new scientific approach in which students of nature should (among other things) avoid the influence of preformed ideas; instead of seeing what they believed, they should believe what they saw.[15] Modern scientists would not disagree with that particular piece of advice.

For a time there were some curious mixtures of old beliefs and new science. In 1600 William Gilbert explained the working of the compass by asserting that the soul of the Earth was a magnet. He also suggested that a magnetic soul was superior to a human soul because it could not be deceived by the senses. William Harvey, who in 1628 finally explained how blood circulates in the human body, believed he was seeing evidence of some vital principle contained within. And, as the King's physician, Harvey was occasionally required to examine suspected witches to see whether they bore the Devil's marks.[16] The French mathematician and philosopher, René Descartes, pursuing his new philosophy as a devout Roman Catholic, held the dualist view of a uniquely human soul. It was a view that he maintained by insisting that non-humans had no minds, could not think or feel as did humans, and that therefore, when dogs were being nailed, fully alert, to dissection boards, they could give only an outward appearance of agony.[17]

Descartes (1596–1650) was a pillar of the scientific revolution. He considered that the human body was a material thing that could be studied by scientific methods, just like any other mechanism. But in believing that thought and reason were located in the non-material mind or soul, he was confident that he could see a clear distinction between what was open to science and what was not. For biology in particular, the scientific revolution was more of a change in ideology than it was a time of great discovery. In biology and medicine the discoveries were to come much later. The scientific revolution was the transition time, when nature was being taken from the realms of scholastic dialogue and exposed to those who specialised in the arts of measuring, testing, poking and probing. Only the human soul, believed to be unique in all nature, was kept and guarded by those who sought other paths to knowledge.

For all our ignorance 400 years ago, European science was in the lead by the standards of the time. By the end of the seventeenth century the new scientific philosophy had prepared the ground for the pursuit and application of science that was to be the mainstay of the industrial revolution. It is a fact, though incidental to our story, that three inventions from China—the magnetic compass, gunpowder and the printing press—were crucially important to Europe's rise to power and influence. But this merely illustrates the difference between technical discoveries and technological progress based on knowing the underlying science. The industrial revolution grew from a uniquely European combination of scientific understanding and social circumstances.[18] Today, science as we know it is practised everywhere, and when new scientific discoveries gain general acceptance it is because they can be replicated by scientists of any background, anywhere in the world.

BEING SCIENTIFIC

I can remember a day in my first two or three years at school when the class teacher told us about fossils. After explaining what they were he told us that he had a fossilised imprint of a leaf on a lump of coal. He took the object from his desk, held it in his hand and had us step up, one by one, to see it. I recall staring at the irregular surface of the coal and not being able to make much of it. Perhaps there was a sort of leafy pattern if you stared hard enough. When we had all been to look he asked if everybody had seen it. We probably all nodded. I don't remember anybody saying no. I certainly didn't—I just felt disappointed. The schoolmaster then turned over his lump of coal and held it up. And there it was: an obvious, perfect, unmistakable imprint of a leaf. He had deliberately shown us the wrong side.

We had just learned, we were told, the first rule of science: 'See what is there to be seen and not what you expect to see', or words to that effect. I felt a fool; we probably all did. It was a good lesson but perhaps not such a clearcut demonstration of gullibility as the teacher thought. There was another variable involved to compound things: some of us might, just *might*, have admitted we couldn't see the leaf if only we hadn't been terrified of the man. Teachers could be very intimidating in those days.

Science has been defined as knowledge that we get from observation and experiment, critically tested, systematised and brought under general principles.[19] No aspect of nature is out of bounds to modern science, and its applications are being extended all the time; but concepts of the supernatural, phenomena outside the influence of natural causes, are by definition outside the scope of science. Science can only work from the assumption that there are general principles to be found, and that causes and effects will have the same sorts of relationship everywhere. Slowly, information is gathered and the connections are made.

Aristotle started from scratch in zoology when he began to collect and describe hundreds of different animals. From his observations he made systematic comparisons and did remarkably well, but he could not possibly have made the right interpretations without a lot more background knowledge. He didn't know, for instance, that blood can be bluish or quite colourless in animals like snails and lobsters, so he thought such animals had no blood. And this raises another obvious but important point: new knowledge in science leads to new questions. Nobody is going to ask why blood comes in different colours if nobody knows that it does.

In the case of blood colour it needed a chemist to show that the blood pigment of snails and lobsters contains copper instead of the iron which occurs in vertebrates and some worms. The pigment with copper turns bluish when it carries oxygen but is colourless when the oxygen has been given up. Nobody could have asked any of that in Galen's day; they were still wondering how blood vessels could carry air. Very often, questions can be precisely formulated but the answers are not forthcoming because other bits of the jigsaw are still missing. When England's William Harvey worked out and quantified human blood circulation he still couldn't demonstrate capillaries. It was another 30-odd years before capillaries and blood cells were seen, by Italian and Dutch microscopists. And after that, more than 100 years passed before oxygen was discovered—by a Swedish chemist in 1772.

Science and technology move in tandem, the one providing tools for the advancement of the other—which leads to better tools. Because knowledge and understanding are always improving, science can never claim to have the final answers. Its descriptions and explanations can only ever be of their time. This is in total contrast to explanations based on the supernatural, which are offered as absolute truths based on authorities and revelations which are to be found by looking backwards, not forwards.

There is no single scientific method, because the techniques of science depend to a large extent on the particular field of study. Biologists may be studying anything from molecules to whales. Animals, of course, are made of molecules, so there are obviously different levels of organisation to be studied. A whale, or a person, has molecules organised into cells; cells into tissues; tissues into organs; organs into individuals; and the individuals are organised into social groups. The traditional scientific approach is to work down to the lower levels in order to get more information—in effect, to pull things to pieces to see how they work. This is known as *reductionism*, and as a way of tackling complex problems it has served us extremely well. Much research in biology is now at the molecular level because that is where the processes occur that most interest the researchers.

But when matter is organised into higher levels of structure it takes on properties that didn't exist before. The flavour of a pudding has not been created until the ingredients are mixed and cooked; you can't find the end product in the ingredients. And you can't appreciate the sweep of a landscape a little bit at a time. Flavours and pictures are emergent phenomena. Scientists must set their sights according to what they want to know; they must ask questions at the levels that seem appropriate and likely to produce answers. When the subjects of study are human beings, however, even the questions can be highly controversial. It is a problem that we shall encounter in later chapters.

NUMBERS AND THEORIES

All animals are incredibly complex compared with inorganic matter, and are correspondingly more difficult to understand. As the British Astronomer Royal, Martin Rees, has put it, 'A frog poses a more daunting scientific challenge than a star'.[20] As well as being complicated, living things vary from one individual to another, and this makes it difficult to draw firm conclusions from small samples which might not be representative. The problem is compounded when you want to compare samples for something specific and it is at its worst when people are involved. Suppose you want to find out if there is any truth in the rumour that women have better memories than men. Which women? What men? Memories of what? Imagine how many variables could affect the outcome of a simple short-term memory test: health, age, interests, education, motivation, cheating, tiredness, alcohol, time of day, test material, attitude of the tester, distractions ... it's obviously not like comparing two bits of metal.

There is a vast literature on how to deal scientifically with complex and variable subjects such as living things. It is a fundamental problem of the life sciences. If we could test or measure all the individuals in a population then at least we would have the true and complete measure of it. But this is rarely possible, so samples have to be taken. With laboratory animals it is common practice to breed closely related strains and to rear the animals under uniform conditions so that variation between individuals is minimised. In future, clones will be used.

With humans and wild animals the variability has to be accepted, and there are a few broad strategies for coping with it. First, if groups are to be compared, they can be matched as far as possible for all the variables that can be identified and might be important. Second, if the numbers are available and costs can be met, the groups can be made big enough for individual variations to be evened out to some extent. In an experimental situation individuals are randomly assigned to experimental and control groups. This is to avoid being unconsciously biased in the selection. For example, the first few individuals to volunteer or be selected might differ in some consistent way from the ones who had to be roped in at the end, so they will be deliberately mixed through the groups. Finally, there is a battery of statistical techniques for sifting chance variations from the sets of figures that researchers produce from their samples.

By convention, scientists commonly consider that they have shown an effect if the probability (P) of the result being due to chance variations in the figures is no more than 1 in 20 ($P = 0.05$). This is an arbitrary value. Researchers may decide to be more stringent and set the significance level at 1 per cent ($P = 0.01$). This means that if there was nothing other than chance at work, there would be only a 1 per cent chance of obtaining the result. Unfortunately there is then a higher probability of dismissing the result as being not significant when the effect really was caused by something other than chance. In short, the more stringent you try to be, the greater the risk of ignoring something interesting.[21]

Scientists can work only with probabilities, and all too often they get less than compelling results from their efforts. But there is one probability that can be relied on to prevent mistaken conclusions from passing into the mainstream body of accepted knowledge: that is the near-certainty that whenever a scientist publishes something interesting, there will be other scientists who set out to confirm it—or otherwise. In science, as in other pursuits, reputations can be made by bringing others down.

Suppose that clear findings have been obtained and have not been disputed. The findings are accepted as being true as far as can be determined. What next? The next problem is to explain them, to account in some way for the outcome of the research. This can open up all kinds of opportunities for creative thinking. The first tentative explanations will be hypotheses but they may come together into an acceptable theory. A series of experiments in the nineteenth century, for example, led to the hypothesis that tiny, invisible organisms caused decay. Other experiments produced similar hypotheses to account for fermentation and the infection of wounds. These led to a comprehensive theory being developed. Infective germs, according to the theory, must be everywhere: in the air and on virtually all surfaces, including human skin. The germ theory revolutionised surgical procedures though it wasn't immediately accepted by everybody. Surgeons were still operating without masks and caps as late as the 1890s, and Florence Nightingale never did believe the germ theory. She believed that disease came from filth and feared that a belief in airborne germs would undermine her campaign for better hygiene.[22]

In science the word 'theory' is not used in the same way as it is commonly used in everyday conversation. A scientific theory is an *explanation*—an attempt to account for what we can see or measure. It makes no sense to talk of 'just a theory' as if it were nothing more than a suspicion. A good theory will absorb new information and may become stronger and more comprehensive as a result. An inadequate theory, which cannot take new findings on board, may become untenable and have to be replaced by another one. Scientists who can topple a generally accepted theory will make a name for themselves. Scientists who are credited with establishing new ones may become household names for centuries.

In the study of human evolution there are many different categories of evidence; there is relatively little opportunity for experiment, and too much temptation to hypothesise. But there is a single, enduring, comprehensive, deceptively simple theory, and it is inseparable from the name of Charles Darwin.

CHAPTER 2

WHAT IS THIS THING
CALLED LIFE?

R EPORTER BOB HOLMES OF *New Scientist* was being precise
and scientific as he worked through his list of phone
numbers and asked one biologist after another for a definition of
life. When a reporter makes long-distance phone calls in order
to put questions to distinguished biologists it is to be expected
that he will have given the questions some thought and will be
choosing his words carefully, but this Holmes was not seriously
hoping to solve any mysteries. He knew that no two definitions
of life, not even biological definitions, were likely to be quite the
same, and he wanted to prove the point in an article he was
writing.[1]

VITAL SPARKS AND CANDLE FLAMES

It is easy enough to list those characteristics of living things which
distinguish them from dead ones and from never-alive ones such
as rocks. The problem is that when you have completed your list
(breathing, eating, reproducing etc.) you will find that some life
forms do not have all the characteristics, while some non-living
ones have quite a few of them. Crystals, for instance, grow and
make exact copies of themselves. A candle flame comes into existence, grows, moves, uses up oxygen, gives off waste and can be
killed—never to flicker again. But crystals and flames were never
alive. Some bacteria, on the other hand, can shrivel into spores
and lie like specks of dust for centuries, showing no sign of life
at all. Nobody knows how long bacteria can remain alive after
putting themselves on hold in this way. Under special circumstances, such as having been freeze-dried, some types might

survive for many millions of years and be able to resume normal life when conditions become favourable.[2] Other bacteria, even at their most active, show scarcely any evidence of the sort of life that plants or animals experience. It is because of examples like this that definitions of life, based on the kind of lives we are familiar with, just won't do. We need to know, in general terms, what it is about bacterial spores (or moulds or cactus seeds) that cannot be found in candle flames (or crystals or computers).

It is easy to think, as did Aristotle, that there must be some special property, some life force, that is the essence of life. In fact vitalism, as this idea is called, underwent a revival in the eighteenth and nineteenth centuries as a reaction to the purely mechanistic explanations of life that followed the scientific revolution. A German embryologist, Hans Driesch (1867–1941), has been described as vitalism's last great spokesman from the ranks of respected scientists. When he discovered that two halves of a sea-urchin embryo could each grow to be a normal sea-urchin, he felt convinced that there must be some sort of guiding force at work—something that could not be described in terms of blind chemistry.

Today, mainstream biology sees no need for explanations of life based on any special force or property that is unique to living things. The properties of the atoms in a living body are exactly the same as those of identical atoms outside it, and there is a constant interchange across life's boundaries. In any case, notions of life forces or vital sparks were never really explanations. To say that something is alive because of a life force explains nothing at all unless you can say what the life force might be. Of course, there are still plenty of mysteries in biochemistry, but at least the science has a solid foundation to build on. Driesch's observations can be explained. We now know that eggs of different types have different distributions of substances which influence development. The pattern in sea-urchin eggs was mapped out by a Swedish embryologist about 30 years after Driesch reported his own findings. Driesch had been separating cells which shared the same part of the development pattern in the dividing egg. If he had taken cells from different parts of the pattern, he would have produced malformed larvae.

A life-support machine provides the body's cells with what they need to carry out the basic functions of life at the cellular level. What no machine can do is to restore the organisation that existed in the healthy person—a level of organisation that has some 10^{14} cells maintaining themselves in tissues, organs and organ systems which all work cooperatively. Lower levels of organisation are easier to support by artificial means. This is not to say that individual cells are not in themselves enormously complicated. Even the smallest living things process vast amounts of information in the course of assembling and maintaining themselves.

Information processing is not, by itself, an infallible sign of life because computers also handle information. We are getting closer to defining life, however, when we recall that all living things both process information and reproduce themselves. Crystals make copies of themselves but they do it in response to the rigid laws of physics, not by processing information. Computers process information but they cannot, so far, use it to reproduce themselves.

When a living cell divides it does it by following a set of instructions contained within itself, and it passes the instructions, not always without error, on to the new cells. Information is therefore processed and transmitted to the next generation by a mechanism of heredity. It is the presence of a hereditary mechanism which decisively separates the world of the living from that which has never lived. Only the processes of life can put together and pass on a recipe for making replicates of the original. Now, some sort of hereditary mechanism, as we shall see, is a prerequisite for Darwinian evolution, and biologists are generally agreed that the capacity for Darwinian evolution can safely be included in any definition of life as we know it. There are no exceptions and no grey areas in that connection.

All life on Earth also exists in cells—either singly, like bacteria, or in organised systems, such as those familiar to us in plants and animals. The first cells were seen in 1665 by Robert Hooke when he examined a slice of cork under a microscope. He saw a honeycomb pattern and called the compartments *cellulae*—Latin for little rooms. They were empty rooms, because cork is dead tissue in which the cell contents have shrivelled away. Hooke did notice that living plant cells are filled with a juice. With a better microscope he would have been able to see a denser patch in the cells, which is the all-important cell nucleus.

Many of the cells in a human body live for only a short time.

Human liver cells, for example, though they last longer than some other types, must maintain their numbers by growing replacements, and this means that our livers have something in common with the proverbial grandfather's axe: grandad fitted a new shaft, father fitted a new head, and now the son is looking after it. In effect, we get a new liver every two or three months for as long as the copying mechanism keeps producing faithful copies of liver cells. Errors accumulate with age, and if the copying mechanism is damaged the cell divisions can run out of control quite rapidly, a condition we know as cancer.

Life as we know it, then, is a process of enormous complexity which takes place inside cells and includes a mechanism of heredity and the capacity for Darwinian evolution. The trouble is that, when biologists try to define life, they have to bear in mind that life had probably been evolving on Earth for hundreds of millions of years before it left any firm evidence of itself. So what form did life take *before we knew it*? And how did it get started to begin with? It is because of uncertainties in these areas that specialists, struggling with definitions, may still disagree about such technicalities as whether life could exist outside of cells. Perhaps it once did.

LIFE'S INGREDIENTS

All the matter in the universe is made up of atoms. Every atom has a central core, the atomic nucleus, which contains particles that carry a positive electrical charge. The atomic number of an atom refers to the number of these particles (protons). Surrounding the nucleus, and orbiting it like tiny satellites, are much smaller particles called electrons, which have a negative electrical charge. In a normal atom the electrical charges cancel out and the atom as a whole is neutral. The electrons, in their orbits, can be imagined as being in a series of holding patterns, because they are attracted to the opposite charge of the nucleus but repelled by each other. This electromagnetic interaction is a form of energy that is fundamental to the processes of life.

Measured across the outer electrons, atoms are in the order of one 10-millionth of a millimetre in diameter, but most of this is empty space because the dense nucleus is only a fraction of an atom's volume. If you imagine the nucleus to be the size of a fist, the nearest whizzing electrons would be more than a kilometre away. Under ordinary conditions, therefore, atomic nuclei never get close to each other so that matter, including living matter, is mainly space.

A few kinds of atom, such as those of helium, are reluctant to join up with any others, but most atoms will readily combine with other atoms according to very fixed rules. Basically, the combining power of an atom depends on the arrangements of its outer electrons: more stable patterns can be made if adjacent atoms form bonds by sharing or transferring electrons. Atoms tend to find the most stable relationships among themselves and join up to form molecules.

When electrons are transferred they carry energy with them, and the strength of a chemical bond is measured by the energy needed to separate the bonded atoms. Atoms or groups of atoms will gain and lose energy when they gain and lose electrons, but gains and losses of energy also result from changes in the position of electrons. It takes energy to move electrons further from the nucleus with which they are associated, and some energy will be released if electrons are moved to a closer holding pattern. At its most fundamental level, life is the channelling of energy flows, and the chemistry involved is referred to as metabolism.

When atoms of the same atomic number join together they form a pure substance, so there are as many pure substances, or elements, as there are atomic numbers. Ninety-two of them occur naturally but atom-splitting physicists have created a few extra ones. Life has found a use for only 26 elements and just four of them (carbon, hydrogen, oxygen, and nitrogen) make up about 96 per cent of the weight of a human body. All life on Earth is made from the same list of ingredients.

A NECESSARY BIT ABOUT CHEMISTRY

Not surprisingly, the four elements that make up the bulk of our bodies will all combine with one another to form molecules by sharing electrons. These cooperative bonds (known as covalent bonds) are not so strong that they cannot be broken at ordinary temperatures compatible with life. This allows atoms to be re-arranged into more kinds of molecules useful to life, some of which are gases that dissolve in water.

Living things are typically about two-thirds water, which would be called hydrogen oxide if it didn't have a common name. Each water molecule consists of two hydrogen atoms combined with one oxygen atom (H_2O). As Thales recognised, water is a remarkable compound in that it exists as a gas, a liquid and a solid within the temperature range of ordinary life. As a gas (vapour) the molecules are largely independent of each other, whereas in ice they form a strong structural lattice. The chemistry of liquid water can be properly described only in highly technical terms, but it isn't difficult to see why its special properties have largely determined the way that life evolved.

A water molecule as a whole is electrically neutral, but despite its neutrality it has positive poles at the hydrogen atoms and negative poles at the electron concentrations, so it is polar, like a magnet. In water, the opposite poles of adjacent molecules are attracted and form attachments called hydrogen bonds. The connections are weak and fleeting but it is the cumulative effect that counts.

Water is not unique in having polar molecules and forming hydrogen bonds: many solids and gases can have polar molecules. It is also common for substances that normally have neutral atoms to lose electrons from their atoms so that they are no longer neutral. The process is called ionisation and it produces electrically charged atoms or molecules (ions). The radiant energy of a nuclear bomb blast is ionising, but so too, to a much lesser extent, is sunshine.

The existence of polar molecules and ions accounts for the way some substances behave in water. For instance, sugar dissolves in water because sugar molecules are slightly polar and become surrounded by water molecules, attached by hydrogen bonds. This effectively prevents the sugar molecules from getting back together again. Table salt (sodium chloride) is soluble in water because its two elements separate to form positive sodium ions and negative chlorine ions. These ions, being electrically charged, also become enveloped by water molecules and remain in solution. Oil, in contrast, has non-polar molecules with no attraction for water. The water molecules, being attracted to each other, will not be pushed aside by the oil, so the oil molecules remain together, either floating on the surface or adopting the shape with the smallest surface area—a sphere. The significance of instant spheres will become apparent in Chapter 4, where we quickly move on from life's origins.

The chemistry of life is largely the chemistry of water. But if any one of the elements has a leading role by virtue of special properties, it is carbon. Pure carbon exists in a crystal state as diamonds and as graphite (pencil lead). In the atmosphere it occurs in combination with oxygen as carbon dioxide (CO_2)— which is a gas. The important thing about carbon is its bonding ability. Four of its six electrons are available for bonding, and they are arranged in a way that permits carbon atoms to form rings or chains to which other molecules can be attached. There are more carbon compounds than there are compounds of all other elements put together.

Complex, three-dimensional structures can be created in which strings of molecules are joined up to make giant molecules known as polymers, but it requires energy to build up big carbon-based molecules, and until the middle of the nineteenth century it was thought that only living things could do it. That is why the study of complex carbon compounds became known as organic chemistry. These days, not only is it possible to manufacture many of nature's carbon compounds but thousands of additional ones have been synthesised. The entire plastics industry is based on them.

Natural carbon polymers make up the familiar tissues of living matter. They are classified into four categories: proteins, carbohydrates, lipids (which includes fats) and nucleic acids (which includes DNA). These four groups are chemically distinct in ways that are beyond the scope of this book, but before moving on there is just one more point that I think should be made.

Nearly all the chemical reactions involved in life are facilitated by a group of proteins called enzymes. Chemically they function as catalysts, substances which prepare others to react but don't get used up in the process. Enzymes in yeast, for instance, catalyse the conversion of sugar into alcohol, a blessing once attributed to the god Bacchus. Enzymes are involved in turning blood into clots and milk into cheese. Enzymes enable life to build up the big molecules of the food we eat, and when we have eaten the food enzymes enable us to break the molecules down again to release the energy in their chemical bonds. Enzymes are very specific in what they do, so it isn't surprising that there are many thousands of them. Specialists have stated categorically that every chemical reaction in biology is promoted by a specific catalyst. They say 'promote' because it is possible for the reaction to occur without the catalyst, but it would be a rare event as it would depend on the right molecules coming together by chance and making a bond, like two jigsaw pieces shaken in a box. Instead, a catalyst acts as a molecular docking station which binds to both bonding molecules and fits them together. Similarly, catalysts help molecules to break apart by binding across the bonds and stressing them.

Until about 20 years ago there was no way to explain how the earliest forms of life could have existed without special proteins to act as catalysts. But there is another possibility, as we shall see in Chapter 3.

NUCLEIC ACIDS AND THE LANGUAGE OF HEREDITY

Anybody who wants a detailed description of the hereditary mechanism has a wide choice of published material to turn to—from great technical tomes and specialist texts through to school biology books and more popular accounts. The DNA molecule has been referred to as the icon of the twentieth century, and this big natural polymer now supports a whole new industry of its own—genetic engineering.

Modern icon or not, nucleic acid was actually discovered in 1869 by a Swiss chemist called Friedrich Miescher. He isolated a substance containing phosphorus and nitrogen from white blood cells and called it nuclein because it seemed to come from the nuclei of the cells. A few years later nuclein was discovered to be slightly acidic and became known as nucleic acid. Nobody then had any idea what it was for.

Before the end of the nineteenth century, separate lines of research had come together. It was known that the little objects within cell nuclei, called chromosomes, were passed on in sperm and egg during fertilisation. It had even been established that the egg and sperm must carry only half the chromosome number so that the new embryo would not receive a double complement. By the end of the century, mainly due to the work of German biologists, it was accepted that the hereditary material was somehow carried by the chromosomes. But just what it was, and how it might work, would not be made clear for another 50 years. Chromosomes contain both protein and nucleic acid, and there was no discernible way by which either could convey the instructions for making so much as a microbe.

Step by step, attention was drawn to the nucleic acid and, around the 1950s, the finale came with something of a rush. At London University, Rosalind Franklin, a biophysicist working in the laboratory headed by Maurice Wilkins, managed to get an X-ray diffraction photograph of DNA. It was this photograph that gave James Watson and Francis Crick, at the Cavendish Laboratory in Cambridge, the clue to the helical structure of the DNA molecule. They worked it out and, together with Maurice Wilkins, collected the Nobel Prize in 1962. Rosalind Franklin, still in her 30s, died of leukemia in 1958.

Two kinds of nucleic acid proved to be involved in the hereditary process: DNA (deoxyribonucleic acid) and RNA (ribonucleic acid). Both are polymers built up of millions of subunits called nucleotides. DNA molecules can be several centimetres long but it would take about 50 000 of them to equal the thickness of a human hair.

Imagine the whole DNA molecule to be a ladder. If it were sawn down its length, through the centre of every rung, each nucleotide would be the half-rung together with its adjoining section of the side of the ladder. The half-rung part of the nucleotide is a base, and these come in four chemical types known by their initials A, G, T and C (adenine, guanine, thymine and cytosine). When our sawn ladder was put back together again (as in the real molecule), the half-rungs would only match up to make the base pairs A–T and G–C. The joins are really hydrogen bonds, and that is the only way they pair off. Imagine the complete ladder being twisted like a winding staircase and you see how its sides form the double helix. Now imagine the winding staircase being wound up beyond recognition, so that it fits into a microscopic space, and you will appreciate the difference between diagrams in books and real life.

So much, for the moment, about DNA. Now a word about proteins. Protein polymers are built up from chains of molecules called polypeptides, which in turn are composed of subunits called amino acids. Twenty different amino acids are commonly found in proteins and some proteins include all of them. Others may have only three, but it is not merely the number of amino acids that determines the type of polypeptide chain: the sequence of them is important as well. To add further to the possibilities, amino acids can be repeated in the sequences, so that proteins often have over 100 amino acid units. I couldn't say in words how many ways there are to arrange 20 units in a chain of 100, but the number is written as 1 followed by 130 zeros, which we are reliably informed is more than the number of atoms in the observable universe.[3] There are specific proteins for the millions of different species, and for different organs of the body and parts of a plant. The human body synthesises about 100 000 different proteins.

The information carried by DNA is a recipe which spells out the types and sequences of amino acids to be built by the new cell. It includes instructions for regulating or modifying the building process during different phases of growth in different parts of the organism. In short, the hereditary mechanism is a complete plan for body-building based on proteins. Because proteins include nearly all the catalysts, a protein recipe can roll out a set of production lines, each with a specialist management team for every type of cell.

Francis Crick and his colleagues, having understood how DNA molecules were structured, spent the next few years working out how DNA molecules were able to store and transmit all this information. At the centre of attention were the four different bases—AGCT—and the possibility that they could form some sort of code. But 20 amino acids could not be represented by only four letters. Neither would pairs of letters be enough ($4^2 = 16$). It had to be triplets of letters ($4^3 - 64$). It was eventually shown that every three letters along the length of the DNA ladder (at least in parts) serve as a triplet, or codon, to specify an amino acid. For example, the sequences GTA and TTC are codons for the amino acids histidine and lysine. Whenever those two codons appear, whether they are in the DNA of babies or beans, the same amino acids are specified.

Once the code had been cracked, a number of laboratories set about working through all 64 permutations of the three bases to determine the codons for every amino acid. It turned out that nearly all amino acids have more than one code name. Histidine, for example, is specified by both GTA and GTG. Even so, not all the 64 possible codons are used up on the amino acids: three triplets are reserved to serve as stops at the end of each run that produces a polypeptide. There are no punctuation marks between the triplets.

In summary, the DNA molecule stores and carries information as if it were punched tape, but with each triplet occupying less than a millionth of a millimetre, one would need many miles of punched tape to transmit a similar quantity of data.

CHROMOSOMES

Only bacteria have their DNA loose in the cell. Other forms of life (leaving viruses out of it for the moment) have their DNA molecules tightly packaged with protein and coiled up in the bundles that we know as chromosomes. These are contained in the cell nucleus. In the living cell the chromosomes bundle up in this way only when the cell is about to divide. After division the chromosomes unwind, and in this extended state the DNA cannot be seen with an ordinary light microscope. The word chromosome (Greek *chroma*, 'colour') is a reference to the way chromosomes can be stained for visibility under the microscope

Each chromosome is a single DNA molecule and there are 46 of them in humans, a number not confirmed until 1956. They can be thought of as 23 different chromosomes with a matching set, except that one pair, the pair that determines sex, doesn't match up as well as the other pairs. The full complement of chromosomes is to be found in every body cell that is capable of dividing.

Stained and seen through a microscope, the chromosomes of a matching pair resemble each other in size and appearance because they are marked off in bands in the same way, like pairs of striped socks. A better analogy would be two print-outs, side by side, with the same headings and blocks of text. At first sight they could be copies of the same print-out but with closer inspection you would find that the text was not always the same under the matching headings. For instance, under the 'eye colour' heading (to use a well-worked example) one chromosome might spell out a sequence for brown while its partner spelled out blue. The partners of each matching pair, of course, come one from each parent.

When a body cell is about to divide, the 46 chromosomes replicate themselves. Enzymes open up the hydrogen bonds between the DNA base pairs so that the ladder rungs break in the middle. This happens in several sections at once. Each half-ladder gets its other half rebuilt so that two ladders are made from one, each consisting of an original string of nucleotides and a reconstructed string. When the whole molecule has been duplicated the chromosomes pull apart and move to opposite ends of the cell before the cell divides between them. In this way one cell becomes two daughter cells and each new cell has all 46 chromosomes. The whole procedure, which is called mitosis, takes about 15–30 minutes per chromosome, so up to ten hours or more for a whole cell.[4]

In the germline cells, that is to say those cells which divide to form sex cells or gametes (sperms and eggs in humans), the chromosomes replicate in the usual way but, instead of pulling apart, the duplicates remain together. The doubled chromosomes then come together in their matching pairs, to give the appearance of four lengths of ladder, side by side. The matching pairs then separate and move to the poles of the cell as in mitosis. But the chromosomes now divide a second time to produce four daughter cells from the original cell. Because there have been two divisions and only one replication, each of the four daughter cells (gametes) will have only half the full complement of chromosomes—one of each matching pair—which is 23 chromosomes in the case of humans. The full complement is restored at fertilisation, when a sperm delivers its genetic material to the egg. This form of cell division, which occurs only in the germline cells, is easy to think of as reduction division. It is a matter of chance which of a pair of chromosomes is included in a gamete, so there are over eight million (2^{23}) different possibilities for any one gamete. Fertilisation squares this number to produce about 70 trillion, so unless we have an identical twin there is virtually no chance of finding that we have a double anywhere in the world.

It often happens, while the matching chromosomes are lying together before the first of the two divisions, that segments of one partner get exchanged with segments from the other in a manoeuvre known as crossing over. This is the genetic reshuffle which mixes up the contributions from each parent so that they are passed on in combination instead of remaining on separate chromosomes. Occasionally the chromosomes don't align themselves properly because a repeated sequence along the DNA molecule gets mistaken for the correctly matching one. Segments of unequal length might then be exchanged so that one chromosome gains a bit and the other one loses. These are not the only kinds of accidental reorganisation that move segments of DNA while the chromosome partners are in physical contact. Because they happen only during the divisions which give rise to the gametes, these recombination errors occur only as a result of sexual reproduction. They have the effect of further increasing the variations that can be passed on in the hereditary recipe. Smaller copying errors, which usually involve only a single base, can happen in any dividing cells when the halves of the DNA molecule are being reconstructed, but more often than not these copying errors are corrected by editing enzymes in a proofreading procedure.

It will be a long time before science is able to explain all the processes that occur between the receipt of a DNA message and the kicking of the prescribed new baby, but the general principles are well understood and an immense amount of detail has been documented during the past half-century. The same basic system is used by all living things. In essence, the cell that receives the genetic message must first read it and then carry out its instructions to make proteins. The reading is done when enzymes open up the hydrogen bonds between the DNA base pairs so that the ladder rungs break in the middle—as happened during replication. At this point, the other nucleic acid I mentioned, RNA, comes into play by acting as a messenger. RNA nucleotides start binding, base by base, to the opened-up sections of DNA. In this way the RNA forms a single strand which, when it comes adrift from the DNA, carries away a working copy of the codon sequences. This whole procedure is called transcription. It is followed by translation, in which the amino acids are put together by molecular machinery called ribosomes, which work along the RNA strand, locking together amino acids in the specified order. This automatically produces the correct polypeptides in the way that makes up the particular protein.

Translation can be very rapid: a chain of 100 amino acids can be put together in as little as three seconds.[5]

GENES

So far I have avoided the word 'gene', but I can't avoid it any longer. It was once thought that genes would be sequences of codons arranged with some regularity along the chromosome. These sets of codons would each determine a particular characteristic such as eye colour, blood group or type of hair. To some extent this is true, but we now know that most characteristics are determined by more than one gene. Many genes may act in concert, and some genes can act as switches to regulate the actions of others, specifying how, when or if they will become active. Genes also occasionally move their position on a chromosome. Nowadays, when geneticists want to be precise in talking about a length of DNA that codes for one particular product, usually a protein, they may call it a cistron rather than a gene.

Much, or most, of a chromosome may not code for anything at all—or at least nothing identifiable. Steve Jones, who is a professor of genetics at University College, London, has described the length of a DNA molecule in terms of a journey through mainland Britain. There are regions of intense productivity, long stretches where nothing seems to be happening, tediously repetitive sectors, and some parts lying in ruins—remnants of the evolutionary past.[6] As little as 10 per cent of human DNA is known to have a coding function.

As more DNA does not necessarily mean more genes, there is not much of a relationship between quantity of DNA and the apparent complexity of the organism. A salamander has about 20 times more DNA than a human being. Nor does the number of chromosomes increase in any regular way with complexity. True, a bacterial cell has only one DNA molecule (actually circular, like a hoop) whereas larger organisms have several lengths of it, but beyond that it is difficult to see much pattern. To compare with the 46 chromosomes of humans, a mosquito has six but a silkworm has 56. A barley plant has 14 but the primitive horsetail plant has 216. A mouse has 40 but a duck has 80. Most organisms have between 10 and 50 chromosomes.

Because genes are functional sequences of DNA bases, perhaps we shall find a more predictable relationship between the relatedness of organisms and the number of their genes. Not many organisms have had all their genes identified but early results have been surprising. A tiny nematode worm has recently had all its genes (its genome) mapped out. The worm is only a millimetre long, has a mere 959 cells in its body, but has 18 000 genes. Specialists working on the worm at Cambridge and Washington universities were sequencing a million bases a month for several years. Progress was posted on the Internet for the benefit of scientists working on other species because there are so many similarities among genes. For instance, a single amino acid difference in one protein was found to account for a change in the movement behaviour of worms in a dish. The same protein in humans is involved in nerve impulses; it is a receptor for a neurotransmitter.[7] It has been thought for some time that humans have at least 80 000 genes in a total of some three billion base pairs, but this may be an overestimate of the genes. The very recent sequencing of human chromosome number 21 revealed fewer than 300 genes, a surprisingly low number.[8] Only 38 000 genes were identified in the 'rough draft' of the genome, which was completed in June 2000. Since then, scientists have been betting on numbers ranging from 40 000 to over 100 000 genes.[9]

The purpose of the Human Genome Project is to determine all the sequences and to map the genes on the chromosomes. We still won't know what all these genes do, nor how they interact with each other. We already know that many of our genes exist in more than one form, as in the different types of gene for eye colour and blood group. Different forms of the same gene are called alleles, and they always occupy the same position (locus) on a particular chromosome. A matching pair of chromosomes will normally have corresponding positions for each gene, so on any locus we might have the same allele twice (both for blue eyes, say) or two different alleles (one for blue, one for brown eyes). No matter how many alleles there may be for a particular gene in a population, as individuals we normally carry only two of them—one from each parent. There are many harmful, even lethal, alleles which we become aware of only when two of the same kind happen to be inherited together. Cystic fibrosis is a well-known example. In the United Kingdom about 1 in 25 people carry one copy of the defective allele. This results in 1 in approximately 2500 children inheriting the defective allele from each parent and being born with the disorder.[10]

For us, as individuals, the lottery of the alleles can be a matter of life and death. But the core of evolution is *replication–variation–selection*, and the alleles are responsible for passing on the variation. Selection falls on the individual. Alleles are the result of genes having been changed in some way. At a minimum the change may involve nothing more than a single base which slipped through the proofreading and editing processes, but whole genes may be affected and recombinations can reshuffle the changes to create much more variation. All permanent changes in a cell's DNA can be broadly referred to as mutations. The word has become familiar enough, but possibly the mutant of science fiction has an undeserved place in public attention, overshadowing the more pervasive influence of mutations in our everyday lives.

In the next chapter we shall see what science has to say about where the raw materials of life originally came from, and how they might have got themselves organised to begin with.

CHAPTER 3

BEGINNINGS

I N 1950 THE BRITISH astronomer, Fred Hoyle, referred to the then controversial 'superdense theory' (which he did not support at the time) as the theory of a 'big bang'. The understatement caught on and it provided a ready label for what has now become the most widely accepted scientific explanation for the origin of the Universe. Since the 1950s a great deal of new evidence has come to light, and Martin Rees is on record as saying he would bet at least 10 to 1 that there was indeed a big bang and that most cosmologists would offer equally strong odds.[1]

COSMIC PRODUCTIONS

In the big bang scenario, a single point of energy became a superdense, infinitely hot ball of radiation which created space and time in an instant of expansion. It was the instant in which the Universe was born, the total amount of universal energy was fixed, and the laws of physics were laid down. Scientific understanding of those laws is now considered to be adequate for explaining cosmic developments back to within the tiniest fraction of a second of this unimaginable event.

Explaining the event itself is the holy grail of theoretical physics because, according to Stephen Hawking and others who have tried to share such thoughts with the rest of us, it is the point at which they must abandon Einstein's theory of relativity and turn to quantum mechanics—and the two theories have yet to be reconciled.[2] Nevertheless, most leaders in this field are confident that they are tackling the right problems. Paul Davies, a well-known theoretical physicist from Adelaide University, has stated that,

In spite of these technical obstacles, one may say quite generally that once space and time are made subject to quantum principles, the possibility immediately arises of space and time 'switching on', or popping into existence, without the need for prior causation, entirely in accordance with the laws of quantum physics.[3]

The various models of the big bang theory have been arrived at by extrapolating backwards from present conditions in the Universe. Three sets of observations are thought to be particularly supportive. First, the Universe is still expanding. Second, it was confirmed by the COBE satellite, launched in 1989, that there is a background radiation of microwaves through the Universe which exactly conforms to the pattern of radiation predicted by the big bang theorists.[4] And third, given that matter is a specialised form of energy, the mix of the light atoms hydrogen and helium is in the proportion to be expected if they had been forged from a fireball which, to begin with, was so hot that no stable atomic nuclei could have existed at all. At trillions of degrees there was only searing light and when, during that first second, energy began to acquire substance, the particles that formed were still unable to bind together because the hot radiation had more energy than the forces that held subatomic particles together.

The first atomic nuclei could form after about three minutes, when the fireball had expanded and cooled to a temperature comparable to that inside the hottest stars—in the order of a thousand million degrees. With further cooling, to a few thousand degrees, protons and electrons would come under the influence of electromagnetic forces so that atoms would begin to form as we know them, with positive nuclei and negative electrons. The COBE findings indicate that around 300 000 years after the big bang there were wispy clouds of matter stretched out over hundreds of millions of light-years. Pulled by their own gravity, these clouds gradually collapsed into denser clusters of galaxies and stars—most of which are made up of about 75 per cent hydrogen and 25 per cent helium. The universe is now about 15 billion years old, roughly three times the age of the Earth.

If there is anything in cosmology that might be described in simple language it is a constantly shining star. It exists because gravity pulls it together until its gases are so compressed and heated up that they resist further compression and begin to radiate heat and light. The sun's white-hot surface is known to be about 6000°C but the interior temperature is put at 15 million degrees. This sort of furnace is a thermonuclear reactor and the reactions are the result of forcing together the protons of hydrogen. It is the energy of nuclear fusion which glows from the sun's surface. For the most part, hydrogen nuclei are being converted into helium.

Bigger stars need to be hotter to balance the stronger gravity, so they burn up their hydrogen more rapidly, thereby having shorter lives. Their lives are more productive, however, in that they can cook up elements that are much heavier than helium. Massive stars are believed to have concentric layers of increasingly heavy elements with cores mainly of iron (atomic number 26). As Martin Rees tells us: 'Not even the centre of the Sun is hot enough to perform these transmutations. But the cores of bright blue stars like those in the Orion nebula, and the intense shocks when they finally explode, can transmute base metals into gold'.[5] Many stars have already exploded as supernovae, scattering themselves into space at thousands of kilometres per second. Others, depending on their size, have come to a more mysterious end. New stars are forming all the time from gas mixed with the dust and debris from stars of the past.

Our sun contains about 2 per cent of elements that came from earlier stars, for the sun is only about five billion years old. As it condensed, it would have rotated more rapidly and been surrounded by a disc of gases, and dust. With the increasing heat of the sun, the lightest elements (hydrogen and helium) would have been cleared away, leaving the dust to form clumps and eventually, through gravity, to amalgamate into planets. Not all authorities agree about the relative roles of dust and larger objects in the formation of planets, but no specialists dispute the conclusion that the Earth, and therefore the raw materials of life, are literally the recycled remains of burnt-out stars. Stardust sounds more poetic.

LIFE'S ORIGINS: EARLY EXPERIMENTS

It is generally accepted that the Earth condensed to a solid body about 4.5 billion years ago. As the outer layers cooled, hot gases burst out from the interior as they do from today's volcanoes. Held in place by gravity, the gases remained around the Earth and became its atmosphere, although it is not likely to have become a stable atmosphere for many millions of years. According to Norman Sleep, a geophysicist at Stanford University, and a number of other authorities, for half a billion years or so there would have been a cosmic barrage of meteorites and asteroids, with some impacts big enough to have repeatedly sterilised the Earth's surface. Large asteroids, up to 270 km in diameter, would have enveloped the Earth in vaporised rock and boiled the oceans, destroying any surface life that might have been present.[6] We shall probably never know whether life existed on Earth during this global holocaust; all that can be said at present is that life has existed continually for the last 3.8 billion years, so in geological terms it was soon established once the planet was reasonably stable.

Until recently, geochemists were of the opinion that, when it did settle down, the early atmosphere was probably a mix of carbon dioxide and nitrogen gas, with some water vapour and probably some combinations of hydrogen with other light elements such as carbon, nitrogen and sulphur. An atmosphere like this, which lacks oxygen but has hydrogen to spare for chemical reactions, is known as a reducing atmosphere (chemically, gaining hydrogen is a form of reduction). In a reducing atmosphere it would be possible for complex carbon molecules to assemble spontaneously, provided there was some activation energy to break existing bonds and get the process started. This cannot happen in today's oxidising atmosphere. Oxygen is a strong attractor of electrons.

Two American scientists, Stanley Miller and Harold Urey from the University of Chicago, published a paper in 1953 in which they described their attempt to simulate the conditions believed to have existed on primitive Earth and so kick-start the chemistry of life. There was nothing very elaborate about the set-up. It consisted of an upper and lower flask connected by glass tubes. The arrangement was sealed except for stopcocks through which samples could be withdrawn. In the upper flask was the primitive 'atmosphere' and in the lower flask was an 'ocean' of hot water. The scientists created 'lightning' in the atmosphere, as the energy of activation, by making sparks jump between two electrodes. When the atmosphere was cooled, droplets of water condensed out and rained into the ocean. After a week of this, the researchers extracted and analysed samples from the ocean. Among many organic molecules which had been created they identified four amino acids—the building blocks of protein. Other scientists were soon doing similar experiments. More than 30 different carbon compounds were produced in this way, including two of the four nucleotide bases (adenine and guanine) of DNA and RNA, but that was as far as it went.

The excitement over amino acids has long since died down but Stanley Miller hasn't given up. Now at the University of California, he reported in 1995 that by raising the concentration of urea in the water (urea was readily formed in the original experiment) it was easy to create the other two nucleotide bases in RNA.[7] Even so, nucleotide bases are not RNA.

More Possibilities

A big part of the kick-start problem is that scientists can only make educated guesses about where, and under what circumstances, life originated. Best guesses about the early atmosphere now point to its having been mainly carbon dioxide and water, with less available hydrogen than was once thought. Better conditions for reduction reactions may have existed underground or on the seabed. Lightning was not the only source of activating energy: cosmic rays, volcanic heat, and even the shock waves from impacting meteors could have been enough to energise the necessary reactions. On the other hand, the sun's ultraviolet rays, unscreened by an ozone layer, would have destroyed nucleic acids. Then again, organic polymers called tholins have turned up in Stanley Miller's flasks—and tholins are natural sunscreens![8] Until something decisive happens, there will be a lot more speculation.

Where on Earth?

The more we learn about existing life, the more possibilities there appear to be for its origins. This does nothing to promote a united research effort, but some fascinating new fronts are being opened up. On the sea bed there are bacteria that thrive near volcanic vents, where it is hotter than the temperature of ordinary boiling water. They get their energy by oxidising hydrogen sulphide, the gas that made school chemistry labs smell like bad eggs. In the Dead Sea there are bacteria with molecular tricks that allow them to do well where other life would be completely dehydrated. Deep in the Earth, living in rock, there are bacteria that need no oxygen at all and have no supply of organic food. They acquire energy by making methane (marsh gas) out of carbon dioxide and hydrogen gas which is present in the rock. From the rocks at the bottom of a South African gold mine, three and a half kilometres below the surface, a bacterium has been identified which doesn't have any particular energy source; instead it specialises in being versatile and can capture energy by chemically reducing a range of different elements, including sulphur—the legendary brimstone. Chemists are understandably keen to find out whether bacteria could even have been involved in producing the gold.[9]

Most of these survival extremists belong to an ancient group of bacteria called archaebacteria. By the late 1970s the work of Carl Woese, a molecular biologist at the University of Illinois, had established this group as a major division of living things. They differ from other bacteria (now called eubacteria) in the chemistry of the cell wall and membrane, and in the structure and function of their genes. Conditions on Earth have changed drastically since life began and what we see among the archaebacteria of today are examples of early metabolic strategies that have survived because suitable conditions have persisted or been re-created—as in volcanic vents, deep rocks, brine pits and sewer pipes.[10]

From once thinking that amino acids were the first step towards life in a primordial pond beneath a reducing atmosphere, we are now faced with such a variety of hypotheses that we are spoilt for choice. Did life first start on the surface, deep underground or on the floor of the ocean? Or independently in all these places? Or did it begin somewhere else altogether?

Not on Earth?

In 1980 the NASA space probe Voyager 1 revealed evidence of organic molecules in Saturn's biggest satellite, Titan. The Giotto spacecraft, six years later, detected masses of organic material as it passed through the dust of Halley's Comet. A meteorite which landed at Murchison, Australia, in 1969 has been found to contain amino acids and hundreds of intriguing microscopic globules.[11]

At NASA's Ames Research Center in California, researchers are now maintaining vacuum-sealed chambers at temperatures of −263°C (10 degrees above absolute zero) containing space gases, dust particles and ultraviolet rays. The gases freeze to the dust grains, and the combination of radiant energy and extreme cold causes molecules to dissociate and rebond in ways that would never happen if the gases were not frozen in place. Many of the resultant chemicals have yet to be named, but in among them have been found tiny droplets just like those from the Murchison meteorite.[12] The significance of this is something we'll return to shortly.

The notion that life could be spread through the Universe by germs is called panspermism, and it has a long history. The serious scientific proposition that life originated somewhere off the Earth has long been championed by Fred Hoyle and his colleague Chandra Wickramasinghe.[13] Writing in the late 1970s these authors tried to persuade us that life forms could not only have tolerated the extremes of space as well as the extremes of a young Earth, they could also have survived the journey between the two environments. From what we now know about the resilience of some microbes, not even the most Earth-centred critics can dismiss this as a daft idea.

Various laboratories have tested bacteria to destruction by subjecting them to environmental extremes. In space-simulating chambers, bacteria have been found to survive the vacuum conditions and intense cold of space, together with the equivalent of 2500 years of UV radiation from starlight. Only one bacterium in a thousand survived, but still there was survival. The possibility of such organisms reaching Earth inside meteorites has to be taken seriously.[14] Some organic precursors of life certainly arrived on Earth from space, probably in substantial quantities. Calculations by Christopher Chyba and the late Carl Sagan of Cornell University show that even today the Earth is collecting about 300 tonnes of organic material each year from space. The stuff arrives on particles less than a thousandth of a millimetre across, which float down instead of hurtling through the atmosphere and burning up.[15]

There are so many possibilities in the search for life's beginnings that when somebody eventually makes headlines with a laboratory success, it may still leave a lot of questions unanswered with regard to the event, or events, of 4–4.5 billion years ago. Everybody agrees that the first type of metabolism could not have needed oxygen, but it is impossible to say just which of the alternative energy paths was first followed. We may never know, and I am not sure that it matters much unless life is found to originate in the freezing desolation of a space chamber instead of in some hot and earthy soup. That would have profound implications for our place in the Universe, because if it can happen in space then it really can happen anywhere—and an infinite number of times.

A Mainstream Position

Leaving the question of metabolism and turning to the other crucial component of life, the hereditary mechanism, we find less scope for disagreement. There is a mainstream position established well enough for dissenters to attract serious attention if they can sound plausible.

I said earlier that it was once impossible to imagine how life, as we understand it, could exist without enzymes to catalyse its chemical reactions. I also mentioned that nearly all enzymes are proteins. Accepting that the present-day hereditary mechanism is always a nucleic acid and that it works by transmitting protein recipes, there is an obvious chicken-and-egg puzzle. Which came first, the protein or the nucleic acid? In the present system neither would be any good without the other. To say that both came into existence at the same time not only compounds the improbabilities, it also fails to solve the puzzle because we are dealing with information transfer here. Information is meaning. Codes have to be recognised and interpreted as information. The necessary mechanism could develop step by step but how could this sort of signalling exist between two kinds of molecules to start with?

A way out of this dilemma was offered in the early 1980s, when Tom Cech of the University of Colorado and Sydney Altman from Yale University found that the RNA molecule can not only carry information, it can act as its own catalyst and replicate itself as well. The replication of a single strand of RNA is mediated by a folded strand of the same molecule, which Cech called a 'ribozyme' from the words ribosome and enzyme. We now also know that single strands of RNA are much less stable, chemically, than the double-stranded DNA. The mutation rate for RNA copying can be a million times higher than that for the high-fidelity replicating of DNA.[16]

With two of the three core elements of evolution (replication and variation) amply demonstrated in the one kind of molecule, it was easy for theorists to visualise an 'RNA world' in which life first consisted of these nucleic acid molecules doing nothing more than lengthening their little chains of nucleotides and replicating. Any mutation which out-replicated the rest would inevitably increase its numbers faster than the others and so become proportionally more numerous. Let us say that one type is converting raw materials to its hereditary line more successfully than are any other types. This selectively advantaged line will eventually predominate, just as selectively disadvantaged ones will disappear if there is competition for the raw materials. This is the principle of natural selection at its simplest and most obvious—a biased game of numbers where the successful competitor has a recognisable advantage and the bias is easy to explain. Change will come inexorably as long as one hereditary line is leaving more replicas than another. And, over time, subsequent mutations will be arising from lines that have already been through the selection sieve; that is how change is accumulated.

Most mutations are disruptive rather than beneficial, so a very high mutation rate, while it could be useful to start with, would probably become counterproductive after a while. Frequent disruptions could begin to outweigh the value of providing new possibilities for selection to work on. This would favour the selection of a more stable molecule, one with a lower mutation rate such as DNA, which differs from RNA in having the double strand and one different nucleotide (thymine instead of uracil). But there are circumstances in which a very high mutation rate will always be an advantage. Viruses are just packets of nucleic acid which can be reproduced only inside more complicated bodies because viruses themselves are incapable of independent life. They can be thought of as bits of DNA or RNA that have escaped control. Our immune systems detect them as invaders and are highly evolved to fight them by molecular binding. But some types, such as the flu virus, keep appearing in different forms because of the high mutation rate of their RNA. The immune system has trouble keeping up with this kind of enemy and finds it difficult to eliminate. Not surprisingly, RNA viruses are among the most troublesome to medical science.

Not everybody is content with the scenario of an RNA world, though it is fair to say that it stands as the mainstream position. Among the dissenting views, perhaps hardest to grasp are the thoughts of Stuart Kauffman, a complexity theorist at the Santa Fe Institute in New Mexico. In complexity theory, a degree of order is created spontaneously when any system reaches a critical level of complexity. Kauffman argues that life is one such degree of order and that it was inevitable, given enough organic complexity to begin with. Mathematics and computer models may support the idea but they fail to impress the test-tube biologists. Among these are scientists, including Carl Woese, who are reluctant to accept that RNA molecules were somehow sustained in an organic soup; they emphasise the importance of first finding the metabolic machinery, insisting that it must have arisen before the RNA molecule.

Research continues, and it will be a triumph for supporters of the RNA world if somebody can show how the molecule can assemble itself and begin replicating without any help. So far, in the test tube, short chains of nucleotides have assembled themselves on clay surfaces, but to get them to grow to the length at which they would act as catalysts for their own replication (50 nucleotides), they had to be given access to a supply of ready-made nucleotides.[17]

Regardless of how we approach the problem, we are stuck with the fact that nobody can ever know the exact mix of molecules and energy that was present when a particular combination came together—and started ticking.

CHAPTER 4

CHEMICALS TO CREATURES

WHATEVER THE FIRST MOLECULES were, there are good reasons for thinking that life began in some sort of cell-like container, rather than floating free in a primordial soup. Inside a capsule the reacting chemicals would be held together and lengthy reactions could be completed before the components became separated.

EARLY CELLS

All living cells, from bacteria to brain cells, are surrounded by the same kind of membrane. It is known as a plasma membrane and is composed of a double sheet of molecules about 6–10 millionths of a millimetre thick. The molecules are lipid—the fatty one of the four categories of carbon polymers listed in Chapter 2. The interesting thing about these molecules is that they are polarised in such a way that they have a 'head' which will bond with water and a 'tail' which won't. When there are enough of them in water the tails spontaneously come together, tip to tip, so that masses of molecules form a double sheet, the massed heads facing the water to form the sheet's surfaces, with the tails to the inside, giving the sheet its thickness. And because a sheet still has exposed edges, it promptly forms a sphere, like a soap bubble. In fact it is very like a soap bubble because it actually remains liquid.

All live cells have a plasma membrane, which is a lipid double layer. And this brings us to the significance of those minuscule bubbles in the Murchison meteorite. They are composed of organic molecules which self-assemble into vesicles and, according to biochemist James Dworkin (who once worked with Stanley Miller), the droplets appearing in the space chambers at the NASA laboratory are made of molecules that behave like lipids. While this doesn't prove anything, it certainly sustains research interest.

Closer to Earth, lipid coats acquired from seawater have been found on water droplets in the atmosphere, and these can become lipid bilayers when the droplets touch down again in the ocean.[1] Whether cells first appeared this way in primal seas remains the big question. Another possibility, if the ingredients were present, is that lipid bilayers accumulated on charged surfaces such as exist on and between microscopically fine layers of clay. Vesicles could then form from a flat sheet by an abstriction process—as in clawing up a handful of tablecloth with outstretched fingers and nipping it off where one's fingers come together.[2] In this scenario, life may have started firmly on Earth not in a soup but on a pizza.

Before we leave the subject of plasma membranes, I should explain that in a real live cell the lipid bilayer is not an entirely waterproof bubble. It is studded with proteins which act like highly specialised pores. These assist the selective passage of molecules and ions into and out of the cell. Depending on the type of cell, some proteins may act as sensors to detect other molecules; as identity tags for the information of other cells; and as physical connections with other cells in the case of tissues.

Plasma membranes are also able to maintain a vitally important electrical potential across their thickness. This is because ions, mainly of dissolved potassium and sodium, are maintained at different concentrations inside and outside the cell. The proteins that selectively allow ions in and out (ion channels) can be specialised for permitting very rapid changes in these concentrations. Equally quickly, this causes the electrical polarity of the membrane to change and, in cells which have this specialisation, every change in polarity triggers off another change further along the membrane. The result is an extremely fast relay of what are technically called action potentials. It is a phenomenon that we all know as a nerve impulse and it permitted the evolution of muscle movement. And brains.

EARLY SEX

Bacteria have various ways of transferring genes from one individual to another. Sometimes fragments of DNA from dead bacteria are taken in by live ones and become incorporated in the bacterial genome. In other cases, bits of DNA can become included in a virus which may then act as a delivery vehicle by invading another bacterium and releasing the DNA. Something very like sexual activity can also occur, even though bacteria are not visibly male and female. In that much-studied bacterium of the gut, *Escherichia coli*, it was found that a fragment of DNA, called a plasmid, can break loose from the main hoop of DNA which lies loose in the bacterial cell. A hollow tube grows from the cell with the plasmid and it connects up with another cell that doesn't have it. The plasmid then replicates and a single strand is passed through the tube into the other cell, where the complementary half of the strand is added. The whole 'sexual' process is called conjugation.

When DNA is moved between individuals there are always possibilities for recombinations. This means there are chances of increasing the diversity that exists among genomes that have already proven themselves in the survival stakes. In evolutionary terms conjugation is an extremely rich source of new material, although most variation among bacteria is still provided by mutations that do not involve recombination.

BACTERIA AND A POINT OF NO RETURN

It was in rocks from Greenland that the earliest traces of life were detected and dated at 3.8 billion years old.[3] There wasn't much to be seen, but fossil bacteria of similar antiquity, and visible with an electron microscope, are known from Australia. Bacteria are all single cells enclosed by a plasma membrane and a tougher cell wall on the outside. They may have their DNA attached to the membrane, but never enclosed within a nucleus. For this reason, bacteria are known to biologists as prokaryotes (Greek *karuon*, kernel). Bacteria come in a variety of shapes but are always very small, typically about a thousandth of a millimetre in diameter. They reproduce by dividing in two, which they can do several times an hour under favourable conditions; and when times are really hard they can form resistant spores and put life on hold for an indeterminate period. Bacteria are an amazingly successful form of life, and for more than 1000 million years they were the only form of life on Earth.

We have seen that some bacteria can channel chemical energy from the most unlikely sources, but about three billion years ago the most innovative metabolism of all time began to evolve. Some prokaryotic cells began to capture energy by channelling the electrons ejected from pigments by the energy of sunlight (one only has to see dyestuffs fade to see this sort of energy at work). Over hundreds of millions of years the energy efficiencies were improved and eventually a type of pigment came to support a complex chemical cycle that could strip the hydrogen from water molecules. Oxygen gas was released as a byproduct and the hydrogen was combined with carbon to make carbohydrates that could be stored as food. The pigment was chlorophyll and the process was photosynthesis.

Things could never be the same again. Life so far had been something of a marginal phenomenon, existing from one minute to the next by channelling the chemical energy that was immediately available. But now cells had a way of channelling the energy into manufactured food and storing it. Unfortunately, the manufacturing process was poisoning the atmosphere with oxygen, which to the forms of life that had evolved without it was a lethal gas.

Some writers refer to this period as the oxygen holocaust, but it wasn't a rapid build-up of oxygen. Judging by the bands of oxidised metals in the Earth, geochemists estimate that it took 400 million years (2.2–1.8 billion years ago) for oxygen to reach its present, stable, concentration of about 21 per cent of the atmosphere. From the start of the build-up, there would have been a dramatic advantage for those organisms that could not only tolerate oxygen but actually use it to gain energy by oxidising food. This is the process of respiration and, needless to say, it could be done only with enzymes.

The first organisms to release oxygen from photosynthesis are believed to have belonged to a group of prokaryotes called cyanobacteria, or blue-green algae, as some scientists still prefer to call them. Today they form a very versatile and widespread group, not all forms of which produce oxygen.

ALL THAT ARE NOT BACTERIA

A new kind of fossil cell appears for the first time in rocks that are about 1.5 billion years old. They are bigger than bacteria, commonly around ten times the diameter, and have evidence of internal membranes. But soft parts of single cells are not easy to find in fossils so we still don't know when the first important changes would have been apparent in real life. From genetic studies some authorities believe that non-bacterial types could have begun to separate from bacterial stock as long as three billion years ago, or even earlier. Non-bacteria are known as eukaryotes (having a proper nucleus) and, whatever form they may have taken, they opened up the evolutionary pathways to the rest of life.

Existing eukaryote cells, as well as being bigger than bacteria, have a cell nucleus bounded by a plasma membrane like that which encloses the whole cell. The chromosomes are located there. The rest of the cell, that is the region surrounding the nucleus, is called the cytoplasm and is partitioned by membranes. It also includes several specialised structures, collectively referred to as organelles. I must mention just two of them because they feature later in our story. In both animal and plant cells there are organelles which handle and store the chemical energy released in respiration. These are compact little bodies called mitochondria and are often described as the powerhouses of the cell.[4] The others I want to name here are the plastids. These are variously shaped structures which contain the chlorophyll in green plants, or the red and brown pigment in seaweeds. In green plants, reasonably enough, the plastids are called chloroplasts.

The problem at this point is how to account for the changes, occurring over an immensely long period, which converted a line of bacterial cells into eukaryotic ones. It took four times longer for the eukaryote cell to evolve from prokaryotes than it did for life to emerge from inanimate chemicals. Every step of the way would have had to confer a selective advantage on the cells involved, otherwise the changes would not have accumulated; they would have come and gone along with the countless other mutations that did nothing to bias the numbers game in their own favour. The fossil record tells us little about the transition, but a plausible scenario can be put together from a combination of fossil evidence and current knowledge of the way that cells work. It is a scenario that has gained general acceptance, at least in outline, and has been neatly summarised by Christian de Duve, who shared a Nobel Prize in 1974 for his work on cells and who has written extensively on the subject since then.[5]

We must start with the assumption that the ancestral eukaryote was a type of cell that needed a supply of organic material as food. Most bacteria (and all animals) are now in that category, and we can refer to them collectively as heterotrophs (Greek, 'fed by others'). Now, heterotroph bacteria do not have any sort of feeding structures. In order to absorb molecules that are too big to pass through their plasma membranes, they simply pour out enzymes which break down the molecules to a manageable size. In other words, they digest their food on the outside. It would obviously improve the feeding efficiency of single cells if they had a highly absorptive surface for soaking up nutrients, but the problem is that the cells also need a strong outer coat for protection. Here we have an uncomplicated example of the way in which design in nature is always a compromise between conflicting needs—in this case, between the needs of absorption and protection. Being the right size is another compromise. Surface areas diminish in relation to volume as size is increased, so it gets harder to service the bulk from the surface. With a sphere, if you double the diameter you will increase the surface area by four (2^2) and the volume by eight (2^3). Ten times the diameter is a thousand times the volume.

Single cells, therefore, *must* be small. Just how small, and with what sort of cell wall, will be determined by finding the most successful compromise between all the conflicting needs—metabolic demands on the surface, cell support, protection, mobility, ease of division and so on. The most successful compromise, among the designs that are made available, will be that which outnumbers the rest in the long term. In view of all the variables, it would be hard to predict a winning design, but with hindsight we can see what kind of changes prevailed.

Here I must digress for just a few lines. On the Western Australian coast there are some curious rock formations that show up at low tide. They look like dark mounds or boulders, scattered over the flats like molehills on a lawn. They are called stromatolites and were formed by bacterial action. Their surfaces are occupied by cyanobacteria but the layers below contain heterotrophic cells, which collect food in the form of waste coming down from the layer above. The whole stromatolite is finely layered because it reflects seasonal growth and it also traps sediments and lays down limestone. The layering can be seen in fossilised stromatolites which have been dated at 3.5 billion years old. We don't know that stromatolites were definitely involved in the evolution of eukaryotes, but they offer a perfect setting in which to visualise the process. End of digression.

When cells are contained in a safe environment with a reliable supply of food, a thick protective coat can become a waste of resources—a handicap in terms of energy efficiency. A colony of heterotrophic cells, safe inside a stromatolite or in some similar situation, might be better off if they lost the outer cell wall and gained a more efficient way of taking in nutrients. With no cell wall, surface area can be enormously increased by folds and pockets in the cell membrane. Cells can then get bigger because the surface area has a way of keeping pace with the increase in cell volume. Some bacterial types have done exactly that.

A flexible outer membrane, which can form pockets, presents another great opportunity. Lipid membranes have a fluidity and capacity for self-healing such that pockets can enclose food particles and trap them by sealing the pocket opening. Food is then contained in a vesicle inside the cell, where it can be digested efficiently with no waste of enzymes. We may now think of big cells, which are capable of engulfing small ones and digesting them internally—a common occurrence among existing eukaryote cells, including our own white blood cells.

In evolution, one thing leads to another. It doesn't need much imagination to see how variations in the folds of a flexible membrane could lead to the network of internal membranes that are found in eukaryote cells. Nor is it hard to visualise how the DNA could have come to be enclosed in a membrane-bound nucleus. If the cell membrane formed a pocket around the DNA, where it lay against the membrane inside the cell, the DNA would become enclosed in a vesicle, just like a food particle but never to be digested. The structure of the cell membranes in modern eukaryotes is consistent with this reasoning.

THOSE ORGANELLES

For a century or so there have been scientists who believed that certain organelles of modern plants and animals are really modified prokaryotes that were taken in by eukaryotic cells and became part and parcel of the higher level of organisation. During the past 30 years or so it has become the mainstream view and is now standard fare in student textbooks. This is largely because of the overwhelming structural and chemical evidence, but it also has a lot to do with the persistence of Professor Lynn Margulis, biologist at the University of Massachusetts, who has been emphasising the role of symbiosis in evolution since the late 1960s. In a narrow sense of the term, symbiosis refers to different species living in close association for mutual benefit.

Chloroplasts were almost certainly free-living photosynthetic cells, and were very probably ancestral cyanobacteria. Today, in all green plants and algae, chloroplasts retain parts of their circular, prokaryote DNA and their bacterial-type membranes and ribosomes. They also divide by splitting, and they do it quite independently of the division of the cell in which they live. But they can no longer live an independent life because they lack essential genes—genes which are now to be found in the cell nucleus. It was a good arrangement: the chloroplasts gained a home and the host cell acquired a food factory. Such things have been shown to happen in the laboratory.

Professor Kwang Jeon, at the University of Tennessee, reported that in the 1960s laboratory amoebae were infected with a bacterium and the two organisms eventually became completely dependent on each other. At first, most of the infected amoebae died and were found to have over 100 000 bacteria inside them. The few surviving amoebae continued to live and multiply, and after a few years they were in a stable association with about 42 000 bacteria in each amoeba. A new symbiotic partnership had evolved, and it has been shown experimentally that after 200 generations (eighteen months) the amoebae cannot survive without the bacteria and the bacteria are unable to live outside the amoeba; each has altered the molecular biology of the other in ways that are still being studied.[6]

Mitochondria, those powerhouse organelles, probably became symbiotic earlier than did the plastids. As they are involved with the energy from respiration, they occur in virtually all eukaryotes. They show the same basic features as the plastids and are thought to have been derived from purple non-sulphur bacteria. Mito-chondria in human cells have been used in evolution studies, and we shall meet them again in this book.

Another organelle story strikes me as the most fascinating of them all. Many eukaryote cells, including those in mammal tissues, have incredibly small hair-like structures attached to them, called cilia. In humans, for instance, millions of them, wafting like grass in the breeze, help to sweep our lungs clean. In a similar way, rows of cilia propel a woman's eggs from her ovaries to the uterus. In men, a single, long version of a cilium functions as the tail on each sperm. When a cilium is cut across and the cut end is examined with an electron microscope, a ring of microtubules can be seen, looking like separate wires in an electrical cable. In a characteristic pattern there are nine double tubules arranged in a ring with another double tubule in the middle. For all animals to have the same pattern, either it must have evolved many times or, more likely, it had a common origin in the very distant past. Where did it come from? The hair-like structures in bacteria were found to be quite different. With one possible exception.

There is a group of bacteria called the spirochaetes, which live in all sorts of places, from mud to the guts of desert termites. They have a twisted shape and can swim with a curious corkscrew action. As disease organisms in humans they cause syphilis. As free-living types they have been seen (and photographed) attaching themselves to larger cells and pushing them along. The spirochaetes feed on leftovers from the larger cell and the latter gets pushed through its food supply. Some spirochaetes have been found to possess fine filaments composed of the same protein that forms microtubules in eukaryotes. It is tempting to say that some ancestral spirochaete must have become symbiotic in eukaryotes at a very early stage, but it will take a lot more research before most biologists are as confident about that as they are about the origins of mitochondria and plastids. But if convincing evidence is ever found, it will come as no surprise to Lynn Margulis.

MULTICELLS: THE END OF IMMORTALITY

Single-celled organisms are still a supremely successful form of life; they comprise more than half the living biomass of the Earth's surface, and the proportion may be much higher when the weight of the underground types has been properly estimated. The evolution of multicellular forms, less than a billion years ago, actually served to increase the variety of situations in which single cells could live. A human baby, for instance, is soon supporting many more bacterial cells than human ones.[7] Nevertheless, it isn't difficult to see the advantages of being multicellular, and there are some interesting pointers to the way it came about.

It sometimes happens that, after dividing, some free-living eukaryote cells will remain clumped together as a cell colony. Scientists at the University of Wisconsin found that when single-celled green algae were exposed to single-celled predators, most of the algae got eaten. Among the survivors were clumps of cells, each clump having anything from four to a hundred cells which the predator had apparently been unable to devour. After 10–20 generations, a steady state was reached in which most cells were living and reproducing as eight-celled colonies. This was presumably the best compromise between security from predation and surface area for nutrient uptake.[8] Predation, by selecting in favour of cells that formed clumps, could have been a powerful force in the evolution of multicelled forms of life.

There is a vast difference between a bundle of identical cells and the finely tuned organisation of specialised cells which makes up an animal's body. But there are living examples of many stages in between; organisation wasn't achieved all of a sudden. Sponges are considered by the great majority of biologists to be very simple animals. They have a particular vase shape but are really just porous tubes made up of three kinds of cells in three layers. They secrete a fibrous skeleton for support (the familiar bath-sponge part) and they feed by filtering particles from the water. Water flows in through the pores of the tube and exits through a hole in the top. The water current is maintained by the inner layer of cells, each of which has a wafting cilium.

Sponges reproduce by releasing eggs and sperms. Fertilised eggs divide up into larvae, which can swim because they have the cilia on the outside. Eventually they settle on a firm substrate, turn inside out, and grow up. Sponges obviously have a simple organisation but can be identified as individual bodies made up of different cells. And yet the organisation is still flexible; there are simple tissues, but no organs. A sponge can be passed through a sieve and the cells and groups of cells can reassemble themselves into the original shape. Fragments of sponge can also grow into whole individuals. Individuality can be shared.

Organ systems arose step by step. Jellyfish have a gut cavity with a single opening, but that is better than none at all. They also have a ring of muscles and a nerve net without any central control—nothing that serves as a brain. Flatworms must also pass food in and waste out through the same opening, but they have right and left sides and a definite head-end for the nervous system. The whole animal kingdom can be described in terms of a few body plans.

Among cells that are living together as an organised individual, there are obvious advantages to be gained from specialisation. Division of labour, either at cellular level or in societies, leads to greater energy efficiencies. With complex jobs it is more efficient to have cooperating specialists than to employ everybody as a Jack-of-all-trades. Bear in mind that when multicellular organisms are reproducing by means of sperm and eggs, those gametes must be carrying a recipe for the whole functional being. And it is the performance of the whole being, in terms of success in passing on more copies of the recipe, that will stand or fall in life.

Cellular organisation lifted the limits on size and opened up the possibilities for the countless shapes and sizes that plants and animals have exhibited. All can be seen as variations on basic patterns which had arisen by the end of the Cambrian Period, some 500 million years ago. Once the eukaryotic cell evolved, a major threshold had been crossed and existing animals, plants and fungi are the current successes from different starting points.

In the multicelled animal line, delivery systems could link the surface with an increasingly distant body centre. Respiratory, digestive and circulatory systems could service every cell. Skeletal and muscular systems could provide support and movement for shapes and sizes that had never before been possible. Nervous systems could provide rapid communication between body parts and improved sensory contact with the outside world. Brains could now think.

The downside is death—the end of immortality. In animals, it is no longer possible for the whole organism simply to split in two and move on through life in everlasting sideways generations. Complicated bodies must revert to sending out single cells to pass on their recipes. And, whether or not they succeed in this, the bodies must die because the necessary organisation cannot be maintained indefinitely. Even death itself is built into the genetic program in ways not yet understood. Programmed death, as Lynn

Margulis and her author son have pointed out, is the original sexually transmitted 'disease'.[9]

Of course there are a great many things that we don't yet understand. Animal bodies can have trillions of cells in hundreds of specialised types, and much has still to be learned about how the whole complex is put together and orchestrated. Perhaps the most surprising thing is how much has been learned in a few decades. After all, the details have been accumulating over a period of some 4000 million years. In all that time, through every minute of every day and in all parts of the globe, organisms have been working through the trials of life, recording their successes, atom by atom, in the growing files of genetic code. The record that we now possess has been maintained without a break, otherwise there would be no record. How many tiny changes, how many giant strides, have been weighed in the balance of selection over that time is impossible even to estimate as we begin to study a few genomes in our laboratories. Changes for the worse have no way of leaving their stories in the scroll of life. Complexity, though, should come as no surprise. Considering that a chromosome can be copied in less than 30 minutes and a chain of amino acids can be strung together in three seconds, it would be an even bigger puzzle if life had not produced something exceedingly complicated after four billion years.

CHAPTER 5

DARWIN'S BEST SELLER

S INCE ANCIENT TIMES IT has been recognised that animals could be grouped according to body pattern—vertebrates with a backbone and four limbs, insects with six legs, the spider group with eight legs and so forth. Aristotle took this into account in his attempts at classification, though without suggesting that shared patterns could be the result of different forms having evolved from the same ancestor. An evolutionary interpretation of life did not occur to the ancient Greeks, although there were vague speculations around the margins of the idea. This seems odd in retrospect and is a reminder of how easy it is to keep staring at the obvious without seeing it until somebody points it out. Then, of course, we cannot understand how we missed it.

This is not to imply that nobody had thought of evolution until Charles Darwin came along: the notion that species might have evolved from earlier types, rather than all having been created separately in their existing forms, had been around for a while before that.[1] During the eighteenth century a number of French academics had remarked on the possibility and Darwin's own grandfather, Erasmus, had actually expressed the thought in verse:

> First, forms minute, unseen by spheric glass,
> Move on the mud, or pierce the watery mass.
> These, as successive generations bloom
> New powers acquire and larger limbs assume.[2]

At the beginning of the nineteenth century the French biologist Jean Baptiste de Lamarck developed the idea by explaining how animals could have become better adapted to their different life-styles by the constant effort of trying to meet their needs. In an often-quoted example, he thought giraffes had grown longer necks because generations of them had been stretching upwards to browse tender leaves and, as a result of this exercise, each generation had been able to pass on a slightly more extended neck. Some years later, in 1844, the Scotsman Robert Chambers (who, with his brother, founded the Chambers Dictionary) anonymously published a work called *Vestiges of the Natural History of Creation*. He made out a case for evolution based on what was known of the fossil record, together with assorted evidence from vestigial organs and embryology. He thought the occurrence of deformities at birth showed that change could occur through birth defects. The book offended the theologians and failed to impress the scientists, but it possibly helped to arouse the interest of the general public.

The anonymous author of *Vestiges* was one of more than 30 writers whom Darwin acknowledged as having declared a disbelief in separate acts of creation by the time Darwin went into print on the subject. Evidently, the possibility that species had arisen from earlier, different forms was an idea that many had passed around but which nobody had found a way of planting. Darwin earned his rightful place in history not just by elaborating the idea but by presenting the most comprehensive and convincing explanation of its time, backed by decades of original observation and painstaking work.

No scientist has attracted more attention from writers than Darwin, so there are plenty of good books on his life and labours and I shall not devote much space to biographical details. The main thrust of this chapter is to show how Darwin's theory still underpins modern evolutionary thinking. We shall return briefly to the historical and social context of his work in Chapter 10, when we arrive in Darwin's century.

THE DOCTOR'S SON

Charles Robert Darwin (1809–1882) was born in Shrewsbury, one of six children of Robert Waring Darwin and Susannah, daughter of Josiah Wedgwood of Wedgwood pottery fame. Susannah died when Charles was eight and his eldest sister took on the mothering role. Charles' father was a successful physician in Shrewsbury and sent Charles to Shrewsbury School for a classical education. Charles didn't like it and was not a good pupil. Classics didn't interest him and by the time he was sixteen he must have been making his true interests so blatant that his father felt moved to say to him: 'You care for nothing but shooting, dogs, and rat-catching, and you will be a disgrace to yourself and all your family'.[3] As the good doctor was paying for Charles' education, one can understand how he felt, but he must have retained some hope because he next sent Charles to Edinburgh to study medicine.

Charles was no good as a medical student either, and he left Edinburgh after a couple of years to study at Cambridge, with a view to becoming a clergyman. Theological studies were not a worry to him but his interests remained firmly centred on natural history and countryside sports, and it seems to have been a matter of chance that he didn't end up as another amateur naturalist working as a Victorian country cleric. A botany professor at Cambridge, John Henslow, took an encouraging interest in Darwin's natural history activities, teaching him botany and intro-ducing him to geology. Indirectly, this was to change the course of Darwin's life.

In 1831 the Admiralty was looking for a naturalist to accom-pany Captain Robert Fitzroy on a survey voyage of the southern hemisphere. In those days it was not unusual for a naturalist to be given a berth on a naval survey ship and then, even more than now, recruitments were made through personal contacts. Professor Henslow recommended Darwin for the post.

With his father's eventual agreement, Darwin signed up and sailed from Devonport on HMS *Beagle* on 27 December 1831. It was expected to be a three-year voyage but Darwin did not get back to England until 2 October 1836, when the *Beagle* docked at Falmouth. In those five years, Charles Darwin was transformed from a keen but aimless naturalist into a uniquely experienced young scientist with an impressive capacity for work. Darwin was a detailed observer with broad interests and in the course of his travels he had looked at coral reefs, earthquakes, glaciers and a range of landforms and vegetation types, from the coasts of Australia and New Zealand to the jungles of Brazil and the moun-tain passes of the Andes. Even his 'shooting and rat-catching' abilities proved useful because at every opportunity he and his assistant collected plant and animal specimens to send back to England. The unpromising student had found his niche.

ROCKS AND A WEDDING

On his return to England, Darwin began writing up the journal of his voyages for publication, but the first edition was mainly to do with geology and this needs a little explanation. The prevailing view of the physical world in the early nineteenth century was that the Earth's surface had been shaped and reshaped by a number of violent upheavals and floods. During these catastrophes, life had been largely wiped out and had somehow been replaced by new forms, perhaps from distant regions. It was an old idea but was championed into the nineteenth century by the great French zoologist Georges Cuvier (1769–1832). Cuvier, on the staff of the natural history museum in Paris, had shown, brilliantly, how the fossil creatures in recent rock layers were entirely different from those in deeper, older layers. An age of reptiles had been followed by an age of primitive mammals and then by an age of later mammals such as mammoths and sabre-toothed tigers. There seemed to be no continuity between them so the doctrine of catastrophism was upheld. The last deluge, which Cuvier believed had occurred about five or six thousand years ago, had evidently cleared the way for people and modern domestic animals.[4]

There were those who did not agree with Cuvier. A Scot by the name of James Hutton (1726–1797) had been one of them. Hutton had presented a quite different history of the Earth's surface by arguing that the Earth was far older than anyone thought and had been shaped by slow geological processes (sedimentation, volcanic uplifting and erosion) that were still, imperceptibly, at work. As an opposite view to catastrophism, Hutton's version of events was called uniformitarianism. It didn't immediately catch on and was particularly unpopular with the Church.

But Hutton's work was not entirely ignored. Another Scottish-born geologist, Charles Lyell (1797–1875), revived and developed the uniformitarian theory in his *Principles of Geology*, published in three volumes in 1830–1833. Charles Darwin (on the advice of Professor Henslow) had carried a copy of the first volume with him on the *Beagle* and used it as a sort of touchstone for much of what he saw. It had served to guide his observations and at the same time Darwin had been able to corroborate Lyell's work. As confirmed uniformitarians, Darwin and Lyell were to become lasting friends. So it was geology, before biology, that brought Charles Darwin to scientific eminence and, if he had gone no further than the journal of his voyages, his treatise on coral reefs[5] and related presentations to learned societies, he would still have had a prominent place among the scientists of his time.

In 1839, the year his journal was published, Darwin married his cousin, Emma Wedgwood, and settled in a house in Upper Gower Street, London. It was an eventful year for he was also made a Fellow of the Royal Society—the oldest scientific society in Britain and still one of the most prestigious in the world. By this time, contrary to his belief when he first set sail, Darwin was convinced that species were not 'immutable' but had gradually changed through variation and selection. It was a conclusion that he decided to keep to himself until he had enough evidence to support it. He could not have predicted how long that would take; he was still beavering away at it when his wife completed their family—with her tenth child.

THE MAN AND HIS BOOK

The theory of evolution by natural selection did not come to Darwin in a single flash of insight. Having confirmed that the Earth was very old, he knew there had been time enough for organisms to have gradually changed. He also knew that individuals within a species were not all identical but showed slight variations in size, shape, colour, behaviour and so on. He collected details of this sort, no matter how trivial, both from his own observations and from the findings of others. He was impressed, for example, by the discovery that the main nerves in a scale insect did not follow the same course in all individuals. On the other hand, he had seen for himself that individual differences were not always insignificant. There were sometimes obvious differences between the races or varieties of animals and plants. He was also familiar with the problems of deciding where to draw the lines between races, subspecies and species. In real life, the divisions are not always clearcut.

In places such as the Galapagos and the rugged lands of South America, Darwin had realised how groupings of animals and plants might have become separated by physical barriers so that they could no longer mix with their parent group but would become further changed in their separate ways. And as a pigeon fancier (and member of two London pigeon clubs) he was entirely familiar with the transformations that could be achieved through selective breeding. Fantails, tumblers, pouters and the rest were, he rightly believed, all rock doves to begin with. With regard to dogs, not even Darwin could believe (as we now do) that they had all been derived from the wolf. He had to 'admit (they) have probably descended from several wild species'. Even so, he went on to say, 'Who can believe that animals closely resembling the Italian greyhound, the bloodhound, the bull-dog ... ever existed freely in a state of nature?'. Who indeed?

Finally, there was the evidence of common descent in what Darwin referred to as the 'mutual affinities' of animals—the fact that groups sharing the same body pattern had evidently put the same structures to different uses. 'What can be more curious', Darwin wrote, 'than that the hand of a man, formed for grasping, that of a mole for digging, the leg of a horse, the paddle of the porpoise, and the wing of the bat, should all be constructed on the same pattern, and should include the same bones, in the same relative positions?'. Moreover, it was well known that the embryos of vertebrates look so similar that it is hard to tell a reptile from a mammal in the early stages—when all of them, including human embryos, have gill slits and a tail.

All these, and other lines of evidence for evolution, were eventually written up as chapters in what Darwin (for reasons that I shall get to shortly) described as an 'abstract' but which was in fact that famous work, *The Origin of Species by Means of Natural Selection*, published in 1859.[6] The quotations above are from its first edition. It is easy to see how, with total dedication, one could spend a lifetime accumulating material for such a book. Today, teams of palaeontologists, embryologists, geneticists, taxonomists and specialised biologists of every sort are still turning up information that would have been grist to Darwin's mill. Good theories just keep absorbing new findings.

We know that Darwin was thinking along evolutionary lines by 1837, but he still had no adequate explanation for how the process could work. If there was a sudden insight at any stage, it came in September 1838, when he read Thomas Malthus' *Essay on the Principle of Population*, which had been published 40 years earlier. Malthus, an economist, had argued that human populations, if unchecked, would always grow geometrically, while food production could be increased only arithmetically. Humans were therefore destined to be forever pushing the limits of subsistence and suffering the ravages of war, famine and ill-health. We know from Darwin's earlier notebooks that there would have been no new ideas for him in Malthus, but something about the *Essay* must have clicked a few of his thoughts together.[7] In *The Origin* he described the 'struggle for existence' as follows:

> It is the doctrine of Malthus applied with manifold force to the whole animal and vegetable kingdoms; for in this case there can be no artificial increase of food, and no prudential restraint from marriage. Although some species may now be increasing, more or less rapidly, in numbers, all cannot do so, for the world would not hold them.

Visualising all the varied forms of animals and plants in a constant struggle for survival, Darwin could see that some would fare better than others because, as he put it: 'Under nature, the slightest difference of structure or constitution may well turn the nicely-balanced scale in the struggle for life . . .'. He had arrived at the idea of natural selection, his mechanism of evolution.

In 1842 the Darwin family moved from London to the country house at Down (now Downe), in Kent, where Darwin was to spend the rest of his life. He tentatively put together his ideas on evolution in a letter to Lyell but got such a negative response that he didn't raise it again for another two years. Then, in 1844, Darwin put together a 'sketch' of his theory and showed it to his new friend Joseph Hooker (who followed Hooker senior as director of the Royal Botanic Gardens at Kew). Hooker was unconvinced.

It would be wrong to suggest that Darwin spent the next fifteen years, to the exclusion of everything else, stewing over his theory. It was rather as though establishing the case for evolution became the background to his life. He continued to accumulate and catalogue all the evidence he could find, from correspondence, publications, talks with specialists, personal observations and by experimenting as far as possible in his own home and grounds. But he also spent time with his growing family; attending to his substantial finances (he never knew poverty); coping with chronic ill-health (not experienced until after his voyages); and working more openly on his up-front professional activities. During the 1840s he produced another book on South American geology, and in the early 1850s he turned out a three-volume treatise on barnacles which alone represented about eight years' work and left him hoping never to see a barnacle again.

By the mid-1850s, for a variety of reasons, Darwin judged the time to be right for going public with his theory, and he began writing it up in earnest. He was well into his intended thick book when, in June 1858, a letter arrived which must have more than ruined his day. It was from Alfred Russel Wallace, a self-taught naturalist with whom Darwin had corresponded, and who was then in the Far East earning money by collecting and selling biological specimens. In this letter, however, he was not offering specimens; he was asking Darwin for his and Lyell's opinion on a scientific paper that Wallace was hoping to publish. It was the essence of Darwin's own theory, which Wallace had arrived at quite independently. Appalled at the thought of being pipped at the post after a 20-year marathon, Darwin turned to his friends and asked Lyell and Hooker what he should do.

The upshot was that, at the meeting of the Linnean Society the following month, Darwin presented a summary of his own theory based on the 1844 sketch and some correspondence. He then immediately followed this with a presentation of Wallace's paper. Darwin records feeling uneasy about the arrangement, but when Wallace learned of it he was content to give priority to Darwin. Darwin, acknowledging 'how generous and noble was his disposition', gave credit to Wallace by writing in *The Origin* that 'Mr Wallace ... has arrived at almost exactly the same general conclusions that I have on the origin of species'.

The readings at the Linnean Society caused surprisingly little stir, but with the issue now out in the open it was not difficult for Lyell and Hooker to persuade Darwin to get something more substantial published—and quickly. So it was that Darwin aborted his plan for a massive book and produced what he called his 'abstract'. In the first edition of *The Origin* Darwin referred to his ideas not as the theory of evolution but as 'the theory of descent with modification through natural selection'. He avoided discussing its implications for human origins but simply stated his belief that 'In the distant future ... Light will be thrown on the origin of man and his history'. *The Origin* soon sold its first print run of 1250 copies and another 3000 copies were hurriedly produced. The book went through six editions in all and is still in print.

Darwin wrote several subsequent books, two of which, in the early 1870s, dealt with aspects of human evolution, but by then, as we shall see in Chapter 10, the public fuss about evolution had largely died down. The old man never actually gave up work; his last publication, on the ecology of earthworms, came out in 1881, and he died the following year, at the age of 73.

WHAT DARWIN DIDN'T KNOW

Ignorance, it seems to me, can be categorised like anything else. First, there are some things, like electricity in the Stone Age, that nobody knows about because our species has yet to discover them. This is the sort of ignorance that we are not even aware of. Second, there are all the questions that we know nobody can answer, even though they are sensible questions. And third, there is the great bulk of specialist knowledge that we know others have, even though we personally don't have—like knowing how to solve long equations or do a heart transplant. Darwin did all he could to keep up to date in every field that was relevant to his theory but he missed vital information through ignorance in the second category as well as the first. He cannot be blamed. These days, obscure information can be tracked down more easily than was possible in Victorian times, even though there is now infinitely more of it.

The Origin was the great turning point in biology—the book that transformed a jumble of pointless, unrelated details into an intelligible, cohesive subject with a proper framework for organising new information. The book was successful because it was convincing, but the idea of evolution wasn't new. It was the *mechanism* of evolution, Darwin's theory of natural selection, that was new. And, ironically, that was the part that other scientists of the period found least convincing. Many people had got used to the idea that some sort of evolution had probably occurred but not even Darwin's close friend, Charles Lyell, was able to accept that the whole process was the result of a blind, mechanistic accumulation of small differences. Since the time of Aristotle it had been accepted without question that nature knew where it was going—that there was a purpose in nature. Christians, with their doctrine of teleology, had no doubt that human beings were central to that purpose. If they were to be convinced that human beings were just one of the outcomes of blind, automatic processes, they would need a lot more evidence than Darwin could muster.

A fundamental problem, of course, was that Darwin knew nothing about heredity. He had shown how individuals differed within varieties and species but he literally hadn't a clue how the differences could arise or be passed on to offspring. As he admitted in *The Origin*: 'Our ignorance of the laws of variation is profound. Not in one case out of a hundred can we pretend to assign any reason why this or that part differs, more or less, from the same part in the parents'.

Darwin did not accept Lamarck's idea that changes could result from the strivings of each generation, but he wasn't able to dismiss the idea that characteristics acquired during a lifetime might somehow be passed on to offspring. In *The Origin*, with reference to the pointing behaviour of gun dogs, he had this to say: 'When the first tendency was once displayed, methodical selection and the inherited effects of compulsory training in each successive generation would soon complete the work . . .'. In fact, pups can't inherit their parents' training any more than human babies can inherit their parents' education.

A related problem was that inheritance was thought to operate by some sort of blending mechanism. A tall and short parent were expected to produce offspring of medium height, just as black and white were expected to produce some shade in between. It was pointed out to Darwin (by an engineer) that if heredity worked that way, all the precious variations for Darwin's natural selection would be blended away, not selected. It was a criticism that must have worried Darwin deeply for he had no answer to it. Darwin later came up with his own inheritance hypothesis, which he called pangenesis. It was supposed to work through particles called gemmules that were made by all cells in response to a body's needs. The gemmules somehow got into the sex cells and carried adaptations to the next generation. The idea didn't help Darwin and didn't even counter the blending problem.

When Darwin wrote *The Origin*, nobody knew any better, but in the later editions, and in other publications, he could have disposed of the blending problem if only he had known about the work of a certain Austrian monk. Gregor Mendel (1822–1884), whose experiments on pea plants established the first principles of genetics, had started his pea studies about three years before *The Origin* was first published. In a monastery at Brünn (now Brno in the Czech Republic), Mendel worked continuously through seven years and about 10 000 pea plants, but reported his results only to the local natural science society. The publication caused not a ripple, and although copies of it were sent to some European and American libraries nobody realised their significance there either. Worse still, a botanist at the University of Munich encouraged Mendel to work on hawkweed (*Hieracium*) instead of peas. Mendel did so and wasn't even able to repeat his earlier findings: hawkweed produces seed without fertilisation! One can't help feeling sorry for Mendel, but there are many sadder stories in science. He was made abbot of his monastery in 1868 and after that, although he maintained his interest in botany, he gave priority to the problems of administration.[8]

Mendel's great discoveries were only rediscovered in 1900, when three European botanists, independently searching the literature prior to publishing their own research findings, came across Mendel's papers. There is no need here to go into details of Mendel's work. His experiments consisted of cross-fertilising different varieties of peas and carefully recording the numbers of each variety that appeared in subsequent generations. In this way he was able to establish the principles which are known today as Mendel's laws. Even though he had no idea what the hereditary mechanism was, Mendel established that the visible traits in a plant (e.g. flower colour, seed colour, pod shape) are transmitted by hereditary factors (now known to be genes) which come in different types (now called alleles), and that the factors from each parent do not blend or alter each other ('contaminate' was Mendel's term). Instead, dominant and recessive traits show up, or fail to show up, according to predictable ratios, and they will keep separating out again in accordance with those ratios.

In choosing pea plants to work with, Mendel made an excellent choice of experimental subject, and he was lucky that the seven traits he selected all showed complete dominance of one allele so that he got clear statistical results. Although Mendel's analysis was correct, we now know that genes don't always behave in a simple Mendelian way.[9] Modern biologists refer to the genetic make-up of an individual as its genotype, and the outward appearance of an individual as its phenotype (Greek *phaeno*, 'show'). Very often the traits we see in the phenotype, such as tallness, are influenced by many genes acting together and by the genotype interacting with environmental factors such as temperature, nutrition and exercise. There are other complications. In some cases incomplete dominance can make a phenotype look as if genes have blended; in other cases one allele can have more than one effect on the phenotype. And then there are all the complexities of recombination and other mutations. Genetics has come a long way since Mendel, but he will always be the acknowledged founder of the science.

Another major knowledge gap persisted in Darwin's theory because there was no way of demonstrating the age of the Earth or the fossil beds that had been found. Darwin felt satisfied that there had been enough time for natural selection to account for all existing life, but how long was long enough? He had written in terms of hundreds of millions of years, basing his estimates on erosion rates and other geological processes, but he was vulnerable to criticism from the physicists. In particular, the influential Lord Kelvin (who developed the Kelvin scale of absolute temperature) believed that the sun and the Earth must be cooling and would have been appreciably hotter only one million years earlier. In no way could Kelvin's calculations be reconciled with Darwin's uniformitarian scenario, in which life evolved in tiny steps while geological processes gradually shaped the Earth's surface over vast stretches of time. This was ignorance of the first category, of course, because at that time nobody even suspected the existence of the nuclear energy that is generating heat. If anyone had told Lord Kelvin that the sun was about halfway through a life of 10 billion years, it would have sounded like lunacy.

By the end of the nineteenth century it was widely accepted that evolution must have occurred, but the explanation put forward by Darwin was, if anything, losing ground. Faced with criticisms that he couldn't answer Darwin had lost some of his earlier confidence in natural selection, and he spoiled later editions of *The Origin* by attempting to satisfy both physicists and Lamarckists. Darwin's best effort was his first.

THE NEW SYNTHESIS

After Darwin, perhaps the most influential biologist of the nineteenth century was August Weismann (1834–1914), one of Darwin's most ardent supporters. Weismann was one of the German biologists who, in the late nineteenth century, concluded that chromosomes must be the basis of inheritance. He also totally rejected Lamarck's thesis that features acquired during the life of an animal could be passed on to its offspring. However, it was Weismann who tried to prove the point by amputating the tails of generations of mice to show that the mutilations couldn't be inherited. Lamarck's supporters claimed it wasn't a true test because the mice had not been responding to a need to lose their tails. Both sides were being rather naive, and modern writers like to make the point with reference to circumcision. Regardless of Jewish boys' past feelings on the matter, they are not being born *sans prepuce*. In more useful research Weismann established the principle that sex cells (or, as he called it, the germ plasm) occupied a separate physical line from the body cells of animals. He also saw the importance of sexual reproduction in producing new combinations of features.

By the early decades of the twentieth century, evolutionary biology was split into several camps. The rediscovery of Mendel's work encouraged geneticists to exaggerate the importance of large mutations instead of the selection of small changes as a driving force in evolution. Other biologists retained Lamarckian views, or clung to the idea of a purposeful progression from lower to higher forms. Part of the trouble was that a variety of specialists were bringing a narrow focus to the issue and nobody was getting the bigger picture.

By the 1940s a consensus had been reached through what amounted to a process of mutual education among the leaders of the field. Once the various contributions had been put together, a spate of books came out that gave an overview of the way evolution works.[10] The title of one of them, Julian Huxley's *Evolution:*

The Modern Synthesis, was taken up as a way of describing the new meeting of minds.[11] By today's standards it was a rather coarse-grained picture that emerged, but it completely upheld the core of Darwin's theory. After the synthesis, Darwinism became known as neo-Darwinism.

DECEPTIVE SIMPLICITY

Some quite daunting mathematics is now involved in studying the genetic aspects of evolutionary change, but the basic principles can still be explained in everyday language. That is a remarkable fact and it cannot be said of any of the other great unifying theories of science. Biologists who care about public understanding have cause to rejoice, but there is a downside. Because it seems so simple, a large proportion of the general public, and even some non-biologist academics, think they have got the hang of Darwinian evolution when they obviously haven't.

The most basic misunderstanding concerns the role of chance. There are people (I have met some) who think scientists are telling them that living things evolved by chance alone. That is the sort of misconception that leads to preposterous questions about the length of time it would take for a team of monkeys to write the works of Shakespeare by hammering on a typewriter. Or whether a heap of components, blown about by wind, would ever become assembled into an aircraft. If I thought that evolution rested on eventualities of that kind I wouldn't believe it either. Chance does supply the raw material, the mutations, for natural selection to work on, but selection itself is anything but random: it is the inevitable, statistical outcome of playing a biased numbers game. It is an inexorable force for change. Given time, it will filter out the characteristics that leave fewer copies of themselves, and it will preserve in DNA those characteristics that best contribute to their own future representation. And of course it doesn't work on one characteristic at a time, it works on whole collections of genes, bringing the long-term survivors together so that the most beneficial characteristics, all things considered, end up in the same individuals.

Another common mistake is to think that, because selection doesn't know where it is going, it will act in favour only of immediate gains in reproduction. This misses the point. While it is true that natural selection can operate only in the here-and-now, it is the longer term that counts. Superior mating and breeding performance is no advantage unless the offspring are successfully reared and are able to carry forward the superior traits for future generations. The writer of the following passage has evidently not understood the role of genes and natural selection beyond the immediate mating performance:

> It was once believed that females mated with males that had the highest potential for a superior genetic inheritance for their offspring. But a recent study of female birds has disputed this argument, claiming that female birds take an extra mating partner not for its genes, but because it may be a better parent and offer a better home territory for raising offspring.[12]

This passage implies that superior genes have nothing to do with being a good parent or maintaining a high-quality territory. In fact these are precisely the sorts of features that are genetically influenced and will be selected as long as they contribute to the survival chances of offspring that perpetuate the traits. Imagine a bird that laid ten eggs in each clutch when the average was only five. At first that might look like a great advantage, but if the parents couldn't manage to raise ten chicks and ended up raising fewer than did parents with the average clutch size, then the trait for big clutches would be a bad one and selection would work against it.

Similarly, outsized human babies might seem to be off to a good start but, statistically, babies of average birth weight (3.6 kg) do best—which is why that weight has become the most common. In countries where babies receive modern medical attention, this stabilising effect of selection will be weakened because babies that deviate from the optimum birth weight will still have a good chance of survival and of passing on their genotypes to the next generation. With modern medicine and technology, many genotypes that were formerly at a disadvantage are no longer exposed to the winnowing of natural selection. It is the long term that counts, but because natural selection can work only on the genotypes that are available, it is forever playing catch-up on environmental changes. With modern baby care we have largely removed a selective pressure to conform. In other cases we are creating new evolutionary pressures, usually for other species but often with consequences for ourselves.

The basic principles of neo-Darwinism really are quite simple, but one does need to think about them a little bit. Beyond the basic principles, there are some aspects of natural selection that require rather more discussion. We shall look at some of these in the next chapter.

CHAPTER 6

DARWINIANS WITH COMPUTERS

AUSTRALIA, IT IS OFTEN said, has more than its fair share of venomous creatures. One of them, the tiger snake, probably lived along much of the southern coast near waterways and wetlands until several thousand years ago—when the sea level rose as a result of melting ice. The tiger snakes have not disappeared but they now have a patchy distribution. Many of them survive as isolated populations on what used to be hilltops but are now small islands off the south coast. These snakes present a fascinating case study in evolution.

On the mainland the snakes grow to more than a metre long and have a choice of prey that includes lizards, frogs and small mammals. The island snakes had to make do with whatever prey remained available on the islands, and some of them ended up with nothing to eat except tiny lizards. Terry Schwaner, a biologist now at Southern Utah University, has weighed and measured thousands of island tiger snakes and studied their diets. On islands where the snakes have a variety of prey they grow to roughly the same size as the mainland specimens. But on Roxby Island, off South Australia, the tiger snakes are still entirely dependent on small lizards for food. These snakes remain quite short and weigh only about 200 grams. They can follow lizards into holes to catch them, and they can feed all year round.

Tiger snakes on another island have gone to the opposite extreme. On Chappell Island, in Bass Strait, there are no small native mammals but the island is used as a breeding ground by a colony of shearwaters, which are large sea birds known locally as muttonbirds. For a few weeks in the year there are plump muttonbird chicks sitting in burrows as food for any snake that is both large enough to eat them and capable of surviving for the rest of the year with no food at all. The ubiquitous little lizards would not repay the energy spent by a big snake in trying to catch them. On this island the biggest tiger snakes measure up to two metres long and weigh 10–15 times more than those on Roxby Island. The females tend to store fat rather than build heavy muscle, which they don't really need, and to give birth not annually but every other year to relatively fewer, bigger offspring. Schwaner and his research colleagues believe these characteristics are best explained as adaptations which have developed within the last 20 000 years in response to available food supplies.[1]

EVOLUTION OF SPECIES

Unlike the stabilising selection, which has maintained the optimum birth weight of human babies, the case of the tiger snakes is one of directional selection, which favours individuals that deviate from the old optimum. In this case selection has favoured giants on one island and dwarfs on another. Zoologists already classify them as separate subspecies. Looking to the future, if the isolated snake populations survive for long enough, other genetic changes, not necessarily adaptations, will accumulate.[2] Eventually there may be such a genetic divergence that different types will not recognise each other as mates, or may not be inter-fertile any longer, or can produce only young that are sterile—like mules. On these criteria zoologists would certainly consider them to be separate species. They would say speciation had occurred. In the case of the tiger snakes, Schwaner believes that the Roxby and Chappell types have already diverged to the point where they would be unlikely to mate, because at compatible sizes any pair would be a hopeless match for maturity.

   In theory, a population of animals (or other organisms) does not need to be geographically separated in order to split into two or more species. If animals mate selectively, from choice or any other reason, a population could become separated into discrete mating groups so that subsequent mutations would accumulate in separate gene pools and cause the groups to diverge further. The initial separation, though, has to be effective. Among domestic dogs, great danes and chihuahuas, for obvious physical reasons, are not likely to mate with each other, but they share a gene pool through their matings with dog breeds of intermediate size. In this example, no type is reproductively isolated.

In reality, most vertebrate species have probably arisen where groups have become geographically isolated, usually at the margin of a species' distribution. Speciation is a fascinating topic, but in vertebrates it is difficult to study within the human time frame. It is the sort of real-life problem that the experts now tackle with computer simulations.[3] In fact, computers have proven invaluable for studying evolution because they are particularly good at handling long, repetitive processes. By combining the repetitive maths with graphic images it is possible to watch things evolve on the screen the way we watch flower buds opening on television. The evolution of an eyeball is especially instructive.

EYEBALLS AND SICKLE CELLS

People who are not familiar with the way evolution works might reasonably ask how an eyeball can possibly start evolving. Either you have a useful eye or you don't; how could it get started a bit at a time? What good is a quarter of an eye? Or half a wing or a third of a head for that matter? Questions like this are based on the assumption that, in the past, eyes, wings, heads and so on were always partly finished versions of what we now see, as if evolution knew where it wanted to get to and would finish when it arrived. In reality, organisms have always been the current models of their time, fully functional for the prevailing circumstances. Evolution cannot anticipate future changes.[4] In the case of eyeballs and vision, plenty of creatures in the world do not have what we would consider to be proper eyes, but they give us some clues to the paths that evolution has followed.

Even the simplest creatures can find it an advantage to be able to distinguish light from dark. Amoebas are just moving blobs with a cell nucleus, but some types can respond to light. More active single cells may have a particular spot of light-sensitive pigment; and some multicelled creatures such as earthworms have light-sensitive spots arranged along their bodies to indicate light direction. Eye spots of this sort have nerves leading to the main nerve centres, and it is an advantage if the eye spot is sunk in a depression, with or without a cover, because this improves the direction signal. There are obvious benefits in being able to detect movement as well as intensity and direction of light. It is better still if a depression becomes a cup, with only a pinhole open to the light, because it will then act as a pinhole camera and actually form images. Certain sea creatures have eyes of this sort. It is a further improvement if the covered cup is filled with cellular fluid instead of water from the surroundings. Part of the cellular fluid can stiffen into a lens. This has happened in vertebrates and some molluscs (squid, octopus, cuttlefish), so the similar eyes of squid and people are a result of evolution having converged on the same outcome from different directions. Biologists are generally agreed that eyes have evolved independently many times, though common genes are involved.

If examples of these various kinds of eyes had not been found, zoologists would have to speculate entirely about the stages that might have been involved in eyeball evolution. As it is, people may still wonder if there really has been enough time for tiny, mutational steps to turn a patch of visual pigment into a functioning eye.

The rate at which mutations appear varies not only among species but among genes, because some are more likely to mutate than others. In multicelled animals, mutations typically crop up at about one per 100 000 to one per million gametes (sperms and eggs). Multiply that by the number of breeding individuals in the population and it can amount to a lot of mutations, especially among small animals but also in humans now that there are so many of us.

The advantages provided by small mutational steps, when they are beneficial at all, are calculated as selective values by comparing the numbers of offspring left by the most successful lines with those of less successful lines. Selective differentials of less than 1 per cent can bring about evolutionary change, but selection pressures commonly reach 10 per cent and can exceed 30 per cent in nature. The selection pressure stabilising human birth weight in the UK was 2.7 per cent in the 1930s, falling to about 1.3 per cent in the 1970s as a result of improved medical care.[5]

Getting back to the question of how long it would take to evolve an eye, this is where computers really show their stuff. Two Swedish scientists set up a program to show what small improvements could do to transform a patch of light-sensitive skin into an eyeball. The researchers made very conservative assumptions about how often mutations could arise, and they allowed improvements only in 1 per cent increments—as measured in optical terms by the computer. They also allowed their evolution to work on only one characteristic at a time instead of operating simultaneously on all characteristics, as it would in real life. The eye developed on the screen as a two-dimensional image and the computer calculated its efficiency as a basic optical instrument. The conclusion was that an eye like that of a fish could evolve from flat skin in fewer than 400 thousand generations. For small animals which reach maturity in a year or so, that would be less than half a million years of evolutionary time, so in the time that has been available it could have happened a thousand times over in each evolutionary line.[6] To put the timing in perspective, though, if all our pet hamsters were turning into elephants at that rate, we wouldn't notice a thing.

In the complexities of real life, individuals are carrying genes for all sorts of evolutionary reasons, most of which we are not even aware of. And when we do determine the function of a gene, it isn't always apparent why it should have been preserved. Sometimes a gene seems to be positively harmful and it is a mystery why it remains common in a population. In the case of human sickle cell disorder, for instance, why should many African people have a gene that distorts their blood cells and kills them in childhood? I raise this question because the answer is well known. The gene is lethal only when the recessive allele is inherited from both parents (as in the case of cystic fibrosis). Individuals with two of the same allelles are said to be homozygous for that gene, whereas an individual with two different alleles of a gene is heterozygous for it. Now, if there is an advantage in being heterozygous for a particular gene, it can outweigh any disadvantage of being homozygous for it. In this instance, heterozygous people have a resistance to *falciparum* malaria. Individuals homozygous for the non-harmful (dominant) allele have no such resistance. In parts of Africa with a high malaria risk, more people are saved by the harmful allele than are killed by it. Natural selection therefore preserves the harmful allele in those areas. The incidence of the allele is declining among African Americans because, where there is no malaria, those with the allele are at a selective disadvantage.

The sickle cell story demonstrates how selection weighs the pros and cons of a single trait in a particular environment, but this is actually happening simultaneously for all traits. Outcomes from the distant past are maintained and carried forward in the record. Think of any event, any threat, disease or cataclysm that occurred in the evolutionary history of any living organism, and its ancestors must obviously have survived it. Genes that proved helpful may still be in its DNA, which is what conservation biologists have in mind when they talk about the importance of biological diversity.

SINCE THE MODERN SYNTHESIS

At the genetic level, evolution means change—alteration to the genetic make-up of populations. Migrations can cause this, and so can events such as the break-up of land masses, volcanic eruptions or forest fragmentation, which split populations into smaller groups possessing different samples of the original gene pool. Some mutations change the structure of proteins without any evident consequence; they seem to be selectively neutral, neither a help nor a hindrance in the survival stakes. These neutral changes will not be subject to selection but will still alter gene frequencies. Specialists argue about the importance of neutral changes and historic events in evolution, but everybody agrees that they are not adaptive. Only natural selection can produce adaptations and virtually all biologists regard natural selection as the core mechanism of evolution.

When academics squabble about the finer points of scientific theory, they don't generally raise much dust beyond the labs and the journals. Evolution is different, because there are passionate disagreements about what gets presented to the public. One professor is constantly in the spotlight, having made it his mission in life to explain evolution in the clearest possible terms to anyone who can read or attend a lecture. He writes well, is a master of metaphor and has recently been described by an American reviewer as one of Britain's national treasures. On the other hand, he has no patience with those who cling to supernatural explanations of life, and he doesn't mince his words when expressing opinions on what many people consider to be sensitive issues. His critics are therefore quite numerous and often strident. I am, of course, talking about Richard Dawkins, who holds the Charles Simonyi Chair of Public Understanding of Science at Oxford.

In his several books, Dawkins has explored many aspects of Darwinian evolution, clarifying and interpreting the trickiest concepts with fresh insight and examples. For many people, including some biologists, Dawkins has been an eye-opener. Others seem to regard him as a threat to society.

One of Richard Dawkins' books was written, he tells us, because he was surprised to find that so many people had not realised what a problem there was for science in explaining what looks like purposeful design in nature. More to the point, they were not aware of the 'elegant and beautiful' solution to the problem that Darwin and Wallace had provided. He called this book *The Blind Watchmaker* and it won tremendous acclaim, as well as a literary award, as one of the clearest and most entertaining explanations of evolution ever written for the general reader.[7]

Richard Dawkins sees no reason to depart in any radical way from the basics of neo-Darwinism. Natural selection is responsible for ensuring that organisms are well adapted to their niche. Adaptations are always a compromise because there are always conflicting requirements: one can't simultaneously store energy in massive slabs of fat and remain lean enough to outrun one's enemies. As in the case of the sickle cell disorder, natural selection will arrive at the compromise that leaves the most copies of itself—having weighed all other needs in the balance. Selection will then stabilise at that state of affairs until the balance of needs gets changed and new directional pressures are created. Natural selection is always playing catch-up, and Richard Dawkins follows Darwin in saying that it operates in very small steps over very long periods. Natural selection, then, is slow and adaptive. Nothing to arouse any passions there, at least not in the general reader.

But there is a convoluted battle line where biology meets the social sciences, and because of one of his earlier books (actually his first) Dawkins was already seen by his critics to be wearing the colours of the wrong side. I am not sure whether I can summarise the issues in a few words but, very broadly, Dawkins gets lumped together with those who are committed to the reductionist approach to science that I mentioned in Chapter 1. They prefer to tackle big problems by breaking them down to their smallest components. I can't imagine anyone objecting to that in principle because science largely depends on it. But one must also bear in mind that there are different levels of organisation to be studied. To succeed at the piano you must create the right sound waves by striking finely tuned strings, but music is more than just a series of sound waves. Everybody would agree with that too, so where's the problem?

The problem is that people have different views about the level of organisation that is appropriate for the study of human behaviour. In caricature, an extreme reductionist might see us as bags of genes, thinking and behaving according to our personal genetic programs. A caricature of the opposite extreme might see people as creations of their social and physical environments, who, by virtue of the human mind, are quite beyond the reach of any genetic influence on behaviour. The truth, as I think all scientists would agree, is somewhere in between, but invariably there are scientific disputes, heated arguments and personality clashes whenever anybody tries to fix the line in any particular context.

I said it was a convoluted battle line and so it is, because those who hate reductionists are usually fighting against what they see as genetic determinism—the idea that we are what we are, and no amount of social improvement will change us for the better. It is a short step from believing that reductionists are all determinists to thinking that determinists are all nasty people who don't believe in trying to help the less fortunate through social welfare policies. And they are the sort of undesirables who tend to be elitists, fascists and racists . . . men like Hitler!

As in all battles, the protagonists on both sides are soon making stereotypes of the enemy and slapping appropriate labels on them. With cooler heads, the label-slappers might privately agree that people of all kinds, and of various religions, can be found in both camps. But in battles about ideology it is public perception that matters, and labels, like slogans, are part of the weaponry. In the postwar years genetics was so strongly linked to politics that research on human genetics was starved of funds and regarded with international suspicion, if not hostility. During that period the safest view to hold was the politically correct one, that human beings were shaped entirely by their cultural environment. Any sliding away from that view was to be opposed on all fronts as a dangerous trend. Compared with those days, what we have now is a much smaller war of resistance, but the battle lines are still there and we shall encounter more skirmishes later in the book.

Back now to Richard Dawkins and his books on evolution.

The title of Dawkins' first book, *The Selfish Gene*,[8] might have been deliberately thought up to antagonise some people, but others praised it as a clever metaphor. The book covers many aspects of evolution, but central to the take-away message is the assertion that genes, acting in concert, are at least partly responsible for many characteristics of individuals in a population. Consequently, there will be genes that are preserved or rejected by the action of natural selection on those individuals. Successful genes are those which leave most copies of themselves in bodies of some sort. Seen in this way, bodies are the unwitting vehicles for the evolution of genes. Calling genes 'selfish' is a metaphor for the fact that they must evolve in the direction that best serves their own perpetuation. From this point of view, to the extent that bodies are the outcome of a genetic program, they must be dancing to its tune. And if this is a valid view for bodies, then what about minds?

It is a matter of degree. One could go along with the gist of the last paragraph but still argue for years in sweeping terms about how much influence genes really have in shaping people's bodies or minds. The question of genetic influence in a particular species needs to be answered, point by point, for each characteristic on the basis of whatever evidence is available. Dawkins wasn't trying to answer such questions in *The Selfish Gene*, and he scarcely dealt with humans at all. We don't need to get into that minefield either, not just yet, but we will look a bit harder at reductionist Dawkins' central assertion that the units of selection are those tiny bits of DNA.

CAUTIONARY TALES AND CONFUSIONS

Stephen Jay Gould is the Alexander Agassiz Professor of Zoology, and Curator of Invertebrates in the Museum of Comparative Zoology at Harvard University. He is an exceptionally productive scientist and for decades has provided a column for the monthly *Natural History* magazine. Collections of these essays have been published as popular science books.[9] The writings of Stephen Gould and Richard Dawkins offer such contrasts that they have virtually become a field of study in their own right, but I shall try to concentrate on the points raised in the previous section.

In 1972 Gould, together with Professor Niles Eldredge, a palaeontologist at the American Museum of Natural History, announced what they called their theory of punctuated equilibria. Basically, this states that species have not arisen by gradual, continuous evolution; instead, speciation tends to be a relatively rapid affair which interrupts long periods of evolutionary stasis. This is consistent with the fossil record in which organisms commonly show no obvious change for tens of millions of years and are then replaced by different types quite abruptly. Abruptly, that is, in geological time; we are still talking about tens of thousands of years.

For some reason the Eldredge and Gould paper generated controversy, and more publications followed. The idea itself, which some are pleased to call evolution by jerks,[10] should not have been provocative. In the neo-Darwinian synthesis, evolution moves in small steps, but there is no reason it has to keep moving at the same rate. Nor did Darwin think so. In *The Origin*, he referred to 'this very slow, *intermittent* action of natural selection' (emphasis mine). Eldredge himself, writing about punctuated equilibria in one of his own books, said 'It should have been noncontroversial. It wasn't'.[11] Perhaps the controversy was caused not so much by the content of the Eldredge and Gould paper as by the media hype that accompanied it. In the popular press this point of fine-tuning acquired the status of a rival theory of evolution and an alternative to everything on offer from people like Richard Dawkins. Publicity of that sort is bound to fuel a fuss, and misunderstandings persist.

In 1979 Gould and another Harvard professor, a geneticist called Richard Lewontin, wrote an article with the title 'The spandrels of San Marco and the Panglossian paradigm: a critique of the adaptationist programme'.[12] This time the message was to the effect that not all features are the result of natural selection; they may not be adaptations at all. Stephen Jay Gould is a clever and eloquent writer but, as the title of that now famous essay might suggest, his favoured style can be other than plain and direct. The Panglossian paradigm is a reference to Dr Pangloss, the optimistic philosopher in Voltaire's satire *Candide*. Dr Pangloss, despite crushing evidence to the contrary, persisted in thinking that everything was for the best in the best of all possible worlds. In Gould and Lewontin's view this is the position of those who believe that all the features of living things are adaptations and must therefore be the products of natural selection. They point out that in St Mark's Cathedral, in Venice, there are magnificent mosaics on the triangular intersections (which they call spandrels) that are unavoidably formed where adjacent arches curve away from each other at the top. Looking at these mosaics one might think that they were the main purpose of the construction, rather than embellishments used to fill incidental intrusions of bare masonry. Gould and Lewontin are telling us that not everything we see in nature is an adaptation; it might be just a byproduct, a concomitant of something else—the equivalent of a spandrel.

Most biologists would accept this. Nobody is claiming, as far as I know, that chubby buttocks are an adaptation to sitting on rock ledges. And many features are merely a result of the laws of physics and chemistry. Blood is red and bones are white because of what they are made of, not because selection has favoured the colours. So, yes, even though seemingly useless features are often eventually shown to have adaptive significance, we should be cautious about thinking up adaptive explanations for every detail of nature. Rudyard Kipling's *Just So Stories* offer an explanation for the elephant's trunk, the camel's hump and so on, and if biologists try to account for everything in terms of adaptation when there is no supporting evidence they are at risk of being lumped together with Kipling.[13] So it was perfectly in order for Gould and Lewontin to sound this cautionary note. And those who keep repeating it are entitled to do so.

But again, an issue of fine-tuning has been turned into a big production for the interested public, as if something has been discovered that challenges the fundamentals of evolution. In fact, there is no challenge here at all. Byproducts notwithstanding, evolutionary biologists are agreed that natural selection is the core mechanism of evolution and that natural selection is adaptive. That is the only way it can work. Gould and Lewontin know that as well as anybody else. Why, then, are professional quibbles about emphases blown out of proportion and allowed to interfere with public understanding?

Professor Daniel Dennet, philosopher and Director of Cognitive Studies at Tufts University in Massachusetts, has considered this question at some length. In his hefty book *Darwin's Dangerous*

Idea, Dennet devotes an entire chapter to Stephen Jay Gould, calling him 'the boy who cried wolf'. Dennet says bluntly that Gould's 'influential writings have contributed to a seriously distorted picture of evolutionary biology among both lay people and philosophers and scientists in other fields'. Gould, according to Dennet, 'has announced several different "revolutionary" abridgments to orthodox Darwinism but they all turn out to be false alarms'. As for Gould's motivations, in Dennet's opinion: 'Gould, like eminent evolutionary thinkers before him, has been searching for skyhooks to limit the power of Darwin's dangerous idea'.[14]

And therein lies the nub of it. Stephen Gould, ever-conscious that distortions of Darwinism can be used to justify social evils, is well known as a warrior on the side opposing Dawkins. We shall meet others later. Of course, many people fear the misuse of science and some might wish that atomic power and genetic engineering had remained in the first category of ignorance. But evolutionary theory is seen as a singular threat because it seems so easy to understand and is accessible to everyone. It can be misunderstood and misused not only by dictators and mad scientists but by anybody who can think, and preach, and influence social policy. This is what really concerns Stephen Gould and others in his camp, and what his critics seem to be saying is that he is so quick off the mark in anticipating such possibilities that he makes a show of sabotaging the weapons that might get misused. Conventional politics may have no relevance (Dennet thinks not), but, for those who are wondering, Stephen Gould tells us that he learned his Marxism at his father's knee[15] and Dennet describes himself as an ACLU liberal.[16]

The modern synthesis remains rock-solid, but in all fairness to Gould's supporters they are entitled to ask whether the current emphases would have been the same without his influence over the past three decades. Gould may have contributed to public confusion, but at the professional level he has probably made some theorists a lot more cautious on the opposing side.

So we come back to those selfish genes and ask the question: Does natural selection really work on genes rather than individuals? Predictably, Gould says that Dawkins is wrong, because 'Selection simply cannot see genes and pick among them directly. It must use bodies as an intermediary ... If most genes do not present themselves for review, then they cannot be the unit of selection'.[17] But then Gould resorts to labels, for he goes on to say: 'I think, in short, that the fascination generated by Dawkins' theory arises from some bad habits of Western scientific thought—from attitudes (pardon the jargon) that we call atomism, and reductionism, and determinism'.

The question is partly philosophical but has been addressed precisely by another evolutionary biologist, Mark Ridley, who is now at Oxford University. In his textbook, *Evolution*,[18] Ridley points out that although natural selection operates at the level of individuals, it can only adjust the frequency of entities that are permanent enough in evolutionary time. Bodies cannot be the unit of selection for the simple reason that, no matter how successful they may be, they die. Nor do their entire genomes get selected because complete genomes cannot be contained in sperm or egg. The units of selection are those lengths of DNA that well-adapted bodies keep passing on. Some will be genes with a single function (cistrons), but there will also be longer fragments with complexes of genes. It was to avoid the problem of defining these inherited fragments that Richard Dawkins called them replicators.

I must emphasise that we are talking here about the mechanical process of natural selection, and that value judgements don't come into it. To fully appreciate this point, imagine a group of people who are supremely successful by any criteria that you care to name, biological or otherwise. But they are prevented from breeding. Those individuals will not make the slightest difference to the DNA in their evolutionary line unless they contribute to raising the children of genetically close relatives. Nor would it make any adaptive changes to the hereditary code if, in a world of enthusiastic breeders, a group of lottery winners used the money to raise five times more children than anybody else. If the lottery winners were a random lot then selection would not have been involved. 'The change in gene frequency over time', says Ridley, 'is not just a passive book-keeping record of evolution'.

BEING GOOD: A REAL PROBLEM

Everybody with an interest in wildlife knows that animals sometimes appear to be quite public-spirited; some social creatures even help to raise the young of others. With the more intelligent animals we might be tempted to think they know what's best for the group, but what about social insects? Nobody thinks that ants and termites have the greater good in mind when they defend their colonies to the death, or when they stay celibate and devote themselves to raising the royal young. So why do they do it?

The explanations that first came to mind were based on the idea that natural selection could work in favour of groups. If one group was more successful than another, no matter why, it would naturally prevail in evolutionary terms. One application of this view was worked up into quite a comprehensive theory in the early 1960s by V.C. Wynne-Edwards of the University of Aberdeen.[19] He had studied the ways in which animal populations appeared to regulate their breeding rates so as to avoid eating themselves out of house and home, and it seemed to him that regulatory behaviour of that sort had evolved because it led to more efficient groups. That sounds quite plausible on the face of it: groups do benefit when their members behave appropriately. The trouble is that it still doesn't explain how the behaviour came to be selected.

The units of selection, call them genes or replicators, are contained in bodies. And bodies have a more rapid turnover than groups because populations last longer than individuals. Bodies can also move from one group to another, and self-servers will be in clover if competitors become altruists. For these reasons alone, individual adaptations are likely to predominate as a result of selection. The characteristics of a group will emerge incidentally as selection chooses the members. In fact, regulatory mechanisms such as territorial behaviour and pecking orders can easily be explained in terms of selection for individuals who are trying only to serve their own interests.

This still leaves us with the problem of accounting for those behaviours that benefit the group but can't possibly benefit the individual. It is impossible for the group members who risk their lives for others, or refrain from breeding, to be the same ones who will prove to be most prolific. So if altruism didn't evolve by group selection, how did it evolve? How can altruistic behaviour be explained in terms of selfish genes?

I must say right now that 'altruistic', like 'selfish', is another metaphor. I am not implying that ants and termites have a social conscience but I can't find a neutral word that means 'unknowingly acting for the benefit of others to one's own detriment'. If I could find one, I would use it.

The solution to the puzzle of altruism came in 1963, when William Hamilton, then a PhD student in London, published some mathematical insights into the importance of family relatedness in the transmission of genetic recipes.[20] Kinship, in a word, explains how it is possible for one individual to pass on more of its genes by self-sacrifice than by breeding. Put at its simplest, if an individual dies in saving the lives of three or more siblings (which, on average, possess half of each other's genes), that individual will have preserved more of its genes than were lost on that fatal occasion. If only nephews, nieces or grandchildren were being saved, there would have to be more than four of them in order for the dead altruist to be statistically ahead in the gene count. For first cousins the number would have to be nine or more.

Obviously, real life is vastly more complicated than this, but no matter how statistically difficult it gets, or how complex the social and behavioural scenarios may be, the principle holds good. Genes that favour altruistic behaviour within groups can be preserved by the mathematics of natural selection on individuals, as long as the rules of relatedness are being applied—consciously or otherwise. In short, altruism can appear at the individual level because of self-serving genes at the molecular level.

Hamilton's publications stimulated an enormous amount of research into altruistic behaviour, especially in social insects and social vertebrates. Social insects are those in which castes of specialised workers devote themselves to ensuring the survival of the queen's eggs and the next generation of closely related workers. It is now believed that this state of affairs evolved independently about a dozen times among ants, bees and wasps and twice in termites.[21] Experiments with birds and mammals that breed cooperatively have also shown that behaviour is in accordance with self-serving genes. Kinship is a strong predictor of when and whether an individual will forgo its own interests to help others. Among birds, for instance, certain bee-eaters were found to be a lot less likely to help in the rearing of their half-brothers and -sisters than with full siblings.[22]

Research in behavioural ecology has reached the stage where the broad influence of kinship on behaviour has been more than adequately demonstrated in real life. But arguments about group selection, usually based on computer models, are continuing. The mainstream view seems to hold that selection could theoretically operate on groups in special circumstances, but examples in the wild are hard to find. Some who favour the idea insist that there must be selection between groups because some groups do better than others. This is not disputed. It simply means that, after self-interests have worked themselves out within groups, the *vehicle* of selection moves up a level. Some groups may well go extinct while others thrive. That is not the same as claiming that group selection can explain altruism.

SEX: ANOTHER PROBLEM

Darwin had a lot to say about sex. He didn't need to be an expert to see that a great many male animals are equipped for fighting off sexual rivals. Others have special features, like fancy feathers, which appear to have no practical value unless to appeal to the opposite sex. He compiled quite a dossier on sexual differences and concluded that those animals in which males and females are physically very different tend to have a polygynous mating system. In monogamous species the sexes are usually similar. As a generalisation it still holds good: we don't see the massive bull as consort to a solitary cow, nor the peacock exhausting himself by displaying to the only hen in his life.

The peacock's train is the classic example of sexual selection taken to extremes. Peahens do choose the most impressive displays and so the display genes get selected, but as with everything else there must be a limit. Peacocks are seriously encumbered by long trains, especially in wind and rain, and must be easier prey for predators. Feathers are mostly protein, which carries another high cost in terms of energy. But as long as the mating benefits outweigh the costs, peacocks will keep growing very long trains. Trains are moulted at the end of the mating season, and I have seen the whole thing jettisoned over a period of about 10 days. Presumably, being burdened for the shortest possible time is the best compromise that selection has been able to come up with. The tail, which lies beneath the train, does not lose many feathers at once because the tail is used in flight.

Observations of this sort are no longer very interesting to the theorists. The real puzzle of sex is why it should exist at all. Sex is the one phenomenon that cannot readily be explained in terms of self-serving genes. Plants, and some small animals, can reproduce without sex so that all the mother's genes are passed to the offspring instead of only half of them. Sharing genes through sex halves their chances of selection. Not only that, but because all the offspring of asexual females will also be females (no male chromosomes) they are able to increase in number much more rapidly (other things being equal) than is possible when half the progeny grows up to produce only sperm. Given this double handicap, how have sex genes come out ahead? It is a fact, as we have seen, that sex enormously increases the amount of variation that natural selection has to work on, so sex does speed up evolution. And sexual populations should be more variable and able to cope with changing circumstances. But when you think about it, if the explanation is that selection has favoured sexually breeding groups instead of those which reproduced asexually, then we are back to group selection again. How did it override drawbacks at the individual level?

There are a number of hypotheses to account for the evolution of sex but none is very simple, and as none is entirely convincing either I won't waste precious space on them. In his textbook, *Evolution*, Ridley sets out some of the main arguments and concludes that we shall just have to keep hoping for an early solution to the sex puzzle.

Human beings will be the main focus of the rest of this book. But we shall be looking at ourselves strictly from the vantage points that I have tried to set up in these past six chapters.

CHAPTER 7

SHAPING OUR FAMILY TREE

LISTENING TO THE LONG-RANGE weather forecasts, with talk about prevailing winds and ocean currents, it is easy to forget that there is nothing permanent about the way the world is arranged. Not even diamonds are forever. Land masses move, climates change drastically, vegetation responds to the changes and animals, at the end of the line, must adapt accordingly. In the nineteenth century, scientists were arguing about how fast or how slowly the land had moved up and down. Nobody even suspected in those days that whole continents were still moving sideways.

In the early seventeenth century Francis Bacon had noticed that Africa and South America would fit together like two jigsaw pieces, but nobody knew why. One suggestion was that the two continents had been joined until they were separated by Noah's flood. It wasn't until 1912 that a non-catastrophic explanation was put forward. In that year a German meteorologist called Alfred Wegener (1880–1930) gave a lecture on the idea of continental drift and followed it up with a book that ran to several editions. Throughout the 1920s it seemed as though his ideas would become established, but the critics kept raising questions that neither Wegener nor anyone else could answer and confidence began to wane. By 1940 there were not many supporters left.

Opinions began to change again after World War II as new technology was applied to the problems. It was discovered that, although the entire Earth is a magnet with the poles as we know them, the individual land masses show ancient magnetic patterns which suggest that they once had a different orientation.[1] When the patterns are analysed, taking into account the fact that the poles occasionally reverse themselves so that north becomes south for a time (it last happened about a million years ago), they provide a picture of the way the world used to be at different times. To add to this, the oceanographers discovered some startling new facts about the seabed. It is constantly welling up from the hotter, softer regions below and spreading outwards. We should not think of runny liquids here but of solids which flow in the way that lead pipes have been flowing when they are found to have sagged out of shape in old buildings.

The current understanding is that the Earth's surface consists of about a dozen plates of solidified magma, some of them carrying whole continents and moving very slowly with the convection currents that are stirring the hotter regions underneath. The forces and pressures involved as they move are beyond our imagining; when the leading edge of one of these crustal plates butts against another one it can shove the Earth's surface into a line of ripples higher than any existing mountain range. The leading edge of the plate eventually dips below the one in front and merges again with the hotter, softer magma underneath. At the same time, new material is being added to its following edge because hotter magma is welling up along ridges in the seabed and spreading outwards. It is as if the continents are being carried about on conveyor belts which are sinking at the front and being lengthened at the back. The theory is known as plate tectonics. Present movements are determined by laser beams bounced off satellites or by using radio-wavelength emissions from quasars. The measurements are accurate to within millimetres.[2] One ridge of spreading in the mid-Atlantic is moving the Americas westwards, further away from Europe and Africa by a few centimetres every year. Australia is heading north towards Asia at a similar rate.

In Earth's time the major land masses have been moved around the globe like pieces on a puzzle board. In the Cambrian Period, between about 500 and 600 million years ago, much of today's land was under shallow seas and life was producing fossil-forming shells. The dry lands at that time could only have been bare rock and are thought to have been strung out around the equator. In contrast, by about 270 million years ago, most of the land was in a single great mass running roughly north–south and actually covering the south pole. It is referred to as Pangea.

Pangea eventually began to break up, first into northern and southern supercontinents (Laurasia and Gondwanaland) and then gradually into chunks broadly recognisable as today's continents.[3] Africa and India moved northwards from Gondwanaland to join with Asia, and the foothills of the Himalayas are still rising with the force of India's docking.

MILK, HAIR AND JAWBONES

I would guess that a great many youngsters, given a time machine, would first set off for the age of giant reptiles—the Jurassic and Cretaceous Periods which lasted from about 210 to 65 million years ago. To survive the trip the travellers would have to hide from the meat-eaters and avoid being trodden on by big herbivores, rather like being in a safari park without a car—only a lot worse. And yet there really were mammals around in that time, although not in great numbers if the scarcity of fossils is any indication.

Perhaps it isn't widely appreciated that mammal-like reptiles were doing very well on Earth long *before* the dinosaurs came into their own. They were called therapsids and were generally short-legged and stocky but grew to be the size of lions. They had legs tucked in underneath, not sprawling out at the sides like modern reptiles, and they very probably had fur. Certainly their teeth were differentiated for biting and chewing and were not just reptilian rows of similar-shaped spikes. They are thought to have been warm-blooded, producing their own body heat from food energy, and they possibly produced milk, but that is something we may never know. We do know that their lower jaws were well on the way to being mammalian. It is a defining feature of mammals that they have a lower jaw made up of a single bone on each side. Two other little bones that appear at the rear of reptile jaws have become the hammer and anvil bones of the mammalian middle ear. Through the mammal-like reptiles we can find the stages in between. Jaws, incidentally, are among those skeletal parts for which there is a continuous fossil record all the way back to the origin. The earliest fish had no jaws at all—just a slit of a mouth that led past gill arches which would later provide the evolutionary material for jawbones.

For tens of million of years the mammal-like reptiles were the dominant vertebrates on land. They thrived at a time when winters were long and cold but they failed to compete with the evolving giant reptiles during the warmer periods that followed. Luckily for us, the therapsids did not disappear entirely without trace, although survival for their descendants evidently involved keeping a low profile. By any definition they had become fully mammalian by the mid-Jurassic, say 170 million years ago, but they were all very small, tended to live in trees and, judging by the large eye sockets, they moved about in the dark. Little survivors of this sort, looking like tree-shrews and probably feeding on insects, were all that kept the mammal's evolutionary line from becoming extinct for more than 100 million years—when yet another climate change, probably involving an asteroid impact, finished off the dinosaurs.

ENTER THE PRIMATES

The branch of biology that deals with classification is called taxonomy (Greek *taxis*, arrangement), and when an organism is classified and named its relationships are taken into account in a hierarchy of shared characteristics. At the broad end of the hierarchy it is easy to see that all animals should go together in the animal kingdom, but at the fussy end the relationships will be much less obvious. Should barnacles go in the limpet group or with the prawns? They look more like limpets but are more closely related to prawns. Some specialists emphasise the importance of finding the true evolutionary relationships, but others say that in many cases this may never be possible so we should settle for a careful assessment of all the obvious similarities.

The animal kingdom is divided into about 36 phyla (singular, phylum) on the basis of basic body plans. Animals with backbones are members of the phylum Chordata—loosely known as the vertebrates, though the simplest types have only a rod of cartilage rather than separate vertebrae. Classes come next, and mammals are one of five classes of vertebrates—the others being fish, amphibians, reptiles and birds. To complete the classification system, classes are divided into orders, such as the carnivore and primate orders within the mammal class; and the orders are divided into families, such as the cat and dog families within the carnivore order. Finally, all living things are given two scientific names, as prescribed by the Swedish naturalist Carolus Linnaeus over 200 years ago. The first name tells us the genus in which the organism has been placed. The second name brings us down to the species. Any smaller divisions, such as subspecies or race, will be between populations so similar that they will be able to interbreed if given the opportunity. By convention, the scientific names are written in italics, with the generic name taking a capital letter. For example, *Panthera leo* and *Panthera pardus* (lion and leopard) are two cat species of the genus *Panthera*. Our own genus is *Homo* and there have been many species.

Mammals evolved rapidly during the Tertiary Period, soon diversifying to fill the niches left by the reptiles, and we can justifiably think of them now as the dominant class of land vertebrates, even though three-quarters of the 4100 living species are no bigger than rats. Primates are distinguished by grasping hands with fingernails and toenails rather than paws with claws. The eyes are set in the front of the skull, which provides binocular vision and good depth perception; and the nipples are never in rows but are paired for the usual birth of only one or two young at a time. The milk glands of all mammals, incidentally, resemble sweat glands—from which they are thought to have evolved.

There is no certainty about where, or precisely when, primates first became distinguishable from earlier insect-eating mammals, and there is plenty of scope for disagreement about primate classification. Specialists can easily distinguish fossil primates from other orders but the very earliest types were intermediate in appearance between the undisputed primates, which are found among fossils less than 65 million years old, and some insectivorous mammals, which were contemporaries of the dinosaurs. The first true primates remained very small, and possibly looked something like modern bushbabies.

Many more types of primates have gone extinct than now exist, although there are still some 236 living species. In classifying primates the taxonomists recognise levels of division that are narrower than the steps in the order–family–genus hierarchy. The result is a bothersome arrangement of suborders, infraorders, superfamilies, families and subfamilies. We shall not get involved other than to say that the first broad division is between the anthropoids (also called simians—the monkeys and apes) and the prosimians, which literally means 'before monkeys' and includes those long-snouted primates—the lemurs and lorises. Clearly, human beings are anthropoids and our closest relatives are the great apes. Accordingly, apes and humans are grouped together in the same superfamily (Hominoidea, the hominoids).

At the family level, humans have always been classified separately from the great apes. The family Hominidae (the hominids) was invariably taken to include humans and direct ancestral types only. All the great apes were grouped in a separate family (Pongidae). This has been harder to justify on zoological grounds since the 1980s, when it was discovered that chimps are genetically closer to humans than they are to other apes. Jared Diamond, a professor of physiology at the University of California, made much of this by referring to humans as the 'third chimpanzee' in the title of his popular book.[4]

UP ON TWO LEGS

The earliest creatures classed as apes are known from fossils found in Egypt and dated at around 30 million years old, but many authorities regard these primates as animals that predate the split between apes and monkeys. A better-known group lived in Africa between 23 and 15 million years ago, but even these types were still not clearly specialised as either monkeys or apes as we know them. They probably climbed and walked in trees on all fours because they lacked the flexible wrists, arms and shoulders that enable living apes to swing below branches.[5] Undisputed apes were living in Africa by the middle Miocene (16–10 million years ago) and they occurred in some diversity through Europe and Asia during the next few million years. But much of the world was becoming cooler and drier through the Miocene and by the end of the epoch (five million years ago) most of the Eurasian types were extinct.

The splitting of the hominid line from other apes assuredly took place in Africa, and when primates next dispersed themselves through Europe and Asia they were walking on two legs. Walking upright, or bipedalism, is an obvious characteristic of the human line and seems to have evolved at a very early stage—before other physical trends, such as reduction in front teeth size and brain enlargement, had gone very far. Many mammals can stand and walk on their hind legs for a short time and apes are quite good at it, but it required a suite of changes to the skeleton and associated muscles to make bipedalism the main way of getting about. The spine needs a lower curve to balance the body over the hips and the pelvis and thigh bones must change accordingly. The big, grasping toe of tree-dwelling apes must point forwards like the other toes instead of sticking out at the side like a thumb. These, and many other physical changes, show up in the hominid fossil record at separate intervals, so specimens from different periods can show novel combinations of features that leave experts squabbling for decades about how best to classify them. The consensus is that the human line and the other apes separated from a common stock somewhere between eight and five million years ago.

In 1924 a mine worker in South Africa found a chunk of rock with such intriguing features that he took it to an anatomy professor in Johannesburg. The professor, Raymond Dart, cleaned it up to reveal the fossilised skull of a child. It had a bigger brain and a rounder jaw than any known ape and it had been found embedded in rock that was millions of years old. To enormous public interest and scientific scepticism, Dart claimed that he had found the fabled 'missing link' between humans and apes and called it *Australopithecus africanus*. The cumbersome generic name (pronounced 'australo-pithy-cus') is a mix of Latin and Greek (L. *australis*, southern; Greek *pithekos*, ape) and one wonders whether the choice had anything to do with Dart's Australian origin. The skull, which is referred to as the Taung skull, has been dated at 2.8 million years.

Since the discovery of the Taung skull, about 20 sites in southern and eastern Africa have yielded australopithecine fossils, and they keep on turning up. Scientists no longer doubt the importance of the genus in human evolution, but with so many fossil specimens now assigned to *Australopithecus*, in half a dozen species, the challenge is to sort out the relationships between them and decide what was ancestral to which. They all lived from about 4.2 to 1.5 million years ago and include some stocky (robust) types, with massive skulls and big teeth, and some lightly built (gracile) species. The relationships between and within groups have proven hard to establish.[6] For example, brawny species (*A. robustus*, *A. boisei*) were living less than two million years ago, while a more gracile type (*A. gahri*) but with even bigger molars was recently discovered and firmly dated at 2.5 million years.[7]

The most complete specimen of an australopithecine found to date was unearthed in Ethiopia's Afar Desert in 1974. The shape of the pelvis indicated a female and the positioning and shape of the leg bones showed that she walked upright. She was nicknamed Lucy (from the title of a Beatles song playing at the time) and was scientifically identified as *Australopithecus afarensis*. In life, some 3.6 million years ago, Lucy would have stood about 107 cm tall and weighed around 30 kg, but adult males could have been head and shoulders taller. Other fossils from the same site suggest an average weight of around 50 kg. Lucy had prominent brow ridges and under her sloping forehead her brain was about 400 cm³. In relation to body size, that is a little bigger than the brains of modern apes but much smaller than those of humans.

Many fragmentary fossils assigned to *A. afarensis* have been found in Ethiopia and Tanzania, but the imagination is stirred most vividly not by any bones but by a perfect set of footprints. At Laetoli, in Tanzania, three australopithecines, one of them walking behind another, evidently traipsed across nearly 24 metres of volcanic ash and left prints of their feet as if they had walked over wet cement. The fossilised prints (together with those of an extinct horse) are dated at 3.6 million years and leave no doubt about the walkers' upright posture. Even the individual toes can be made out—all pointing forwards like our own. The prints were uncovered by Mary Leakey's team in 1976.[8]

Nobody can be sure where Lucy's type fits into our family tree, but among various suggested arrangements it is common to see it linked to the first *Homo* through *A. africanus*. The trouble is that *A. africanus* hasn't yet been identified in East Africa where the earliest *Homo* fossils have been found. As more pieces of the puzzle come to light a clearer picture could come into focus. But, for a time, the puzzle might just keep growing.

MISSING LINKS AND BROKEN CHAINS

We have seen that *Australopithecus* and related genera lived in Africa at least 4.5 million years ago and that some were still around as contemporaries of the genus *Homo*. But we still don't know what was ancestral to the australopithecines. Neither can we be sure which, if any, of the known australopithecines was ancestral to *Homo*. These links have yet to be closed. But within *Homo* we have other links that can be put together in different ways—except that they won't make a single long chain because there are lots of dead ends. To change the metaphor, *Homo sapiens*, like many species, is on one twig of a branch of a bushy family tree, and it isn't even clear where the branch ends and the twig begins.

In practice, fossil hominids are assigned to our own genus if they satisfy one or more of four criteria: a brain size of at least 600 cm³; evidence of language (from brain shape revealed on the inside of the cranium); a human grip with opposable thumb; and, finally, the ability to make stone tools. Two anthropologists, Bernard Wood from George Washington University and Mark Collard from London University, have recently argued that none of these criteria is satisfactory and that *Homo* is not a good genus as they define that taxonomic category.[9] Accepted or not, their argument demonstrates the problems of defining our own genus, and, as the next example shows, that could make the difference between handy man and handy ape.

In 1964 Louis Leakey coined a name for a fossil hominid found in Tanzania's celebrated Olduvai Gorge and dated at about 1.8 million years. Leakey decided on the name *Homo habilis* (handy man) because the remains were found together with the earliest type of stone tools, and it was concluded (amid some controversy) that *H. habilis* had made them. Another possibility is that the tools were made by one of the australopithecines that was around at the time. To the ordinary observer there isn't much difference between the fossil types. The first *H. habilis* to be found had a brain of only about 650 cm³, though later finds pushed the size to more than 700 cm³—which is bigger than any recorded for *Australopithecus*. But *H. habilis* was not otherwise distinguished, having a low forehead and projecting face, and standing about 1–1.5 metres tall with an average weight of around 34 kg.

The first deliberately shaped tools do appear at about the same time as the earliest *H. habilis* fossils—some 2.5 million years ago. The *H. habilis* fossils might represent more than one species, but that isn't the only possibility. After a careful assessment of all the characteristics, Wood and Collard recommended reclassifying all *H. habilis* as *Australopithecus*, at least for the time being. The genus *Homo*, in the opinion of these authors, should contain only those species which resemble modern humans more than they resemble the australopithecines and their relatives. Wood and Collard offer half a dozen criteria by which to make the judgement. If *H. habilis* is reclassified as an australopithecine, then on present evidence the human genus was not the first to start making stone tools.[10]

HOMO

After Darwin, there must have been many individuals with a passionate interest in the 'missing link' between people and apes. One of them was Eugène Dubois, a Dutch anatomist and geologist. In 1891, nine years after Darwin's death, Dubois was working as a military surgeon in what was then the Dutch East Indies. He was digging into a hillside in eastern Java when he found part of a fossilised jaw, a skullcap and a thigh bone. Together, they indicated some sort of semi-human that had walked upright, possessed a flattish forehead and had massive brow ridges. Its brain, though small, must have been far bigger than that of any ape. Judging from other fossils in the area, the bones were likely to be several hundred thousand years old. Dubois published his findings and later became a professor of geology at Amsterdam University. Nevertheless, he found it so difficult to persuade the scientific establishment to accept his find as human that he stored his 'Java man' under the floorboards and refused to show it again for nearly 30 years.

Dubois was vindicated in the 1920s when a site in China yielded a total of 14 similar skulls, together with other bones and tools of stone and bone. 'Peking man' lived somewhat later than Java man, was less robust and had a rather bigger brain, but the similarities were obvious. Plaster casts were made and distributed. The precious originals were kept for study at an American-supported medical college in Beijing (then Peking). In 1941, when a Japanese attack seemed imminent, it was planned to ship the fossils to safety in America under US Marine escort. The entire collection was carefully packed up and sent by train to a Marine outpost to await shipment, but the outpost was surrendered to the Japanese before the troopship arrived. The fossils disappeared and their fate remains a mystery—despite an offer of $150 000 for information.[11]

Both Peking man and Java man (they were described under scientific names) are now considered to be examples of *Homo erectus*, the human species which evolved in Africa and spread from that continent to Eurasia more than 1.7 million years ago. Specimens from Africa, including an almost complete skeleton from Kenya known as the Turkana boy, are also commonly assigned to this species, although some authorities separate the earliest types under the name *H. ergaster* and regard them as a more primitive stage that lasted from about 1.9 to 1.5 million years ago.[12]

Nobody denies that when African and Eurasian types are lumped together as *H. erectus*, we have a species that contains a pretty wide variety of physical types. Brain size ranges from around 850 cm³ in African fossils to more than 1000 cm³ in those from Java and China. All have brow ridges but the African skulls are more lightly built. If an animal covers such a wide geographical area for such a long time, how much physical variation is it reasonable to expect within one species? There are no simple rules for answering that question. Some authorities (referred to as 'splitters' in the trade) recognise six or eight species of *Homo*; others (the 'lumpers') are satisfied with only three. Both lumpers and splitters can make reasoned decisions without pretending to know exactly where the lines were drawn in real life.

A CONFUSION OF HUMANS

Just because our genus came out of Africa a lot earlier than anybody once suspected, it doesn't necessarily mean that we are the direct descendants of those early migrants. There are various fossil forms that are intermediate in appearance between *H. erectus* and 'anatomically modern humans' as the anthropologists describe us today. Details are outside the scope of this book so I'll just say that the earliest fossils to be considered modern humans (*Homo sapiens*) are about 100 000 years old; and those intermediate types that are judged to be closer to *H. erectus* than to *H. sapiens* are collectively referred to as 'archaic moderns'. I can also say that sorting out the possible relationships through this level is another highly contentious business.

In Chapter 4, I mentioned the mitochondria, those organelles which retain some of their own DNA and divide quite separately from the body cells in which they exist. Mitochondria are inherited through the egg in sexual reproduction, but the few that might travel in a sperm were thought not to be carried into the egg because only the sperm nucleus gets through to deliver the chromosomes at fertilisation. So until recently it was accepted that a separate genetic line could be followed in mitochondrial DNA through the females of a species. This line would not be affected by the sexual mixing that complicates the history of DNA in the chromosomes. Mitochondrial DNA, like bacterial DNA, replicates without being proofread or edited, so mutations are relatively frequent. Not only that—it was also found that mutations appeared at a fairly constant rate, marking off time like a molecular clock. It was realised that once the mutation rate was known it would be possible to compare living populations and see how much time had passed since they became genetically separated.

In the 1980s, scientists at the University of California at Berkeley, in the lab of the late Alan Wilson, announced that they had traced a line of mitochondrial DNA, from a sample of modern women from many different ethnic groups around the world to a point of convergence in an African female who lived approximately 200 000 years ago. In the popular press an 'African Eve' had been discovered.[13] But, as Richard Dawkins has pointed out, it is mathematically inevitable that if we are able to follow a sample of modern females back through the female line, we must converge on an individual woman who was the most recent common ancestor on the female side. By the same reasoning there must also be a most recent male common ancestor, possibly less distant. We need science only to tell us where and when these common ancestors lived.[14]

Africa still holds up as the place of most recent common ancestry, as well as being the place of human origin, but the timing of our African Eve has taken a battering from more recent genetic studies. Molecular clocks, it seems, are not as reliable as was once thought, and mitochondrial DNA from males can get through and even recombine with that of females to create more havoc with our time-reading.[15]

If all modern humans really had an African common ancestor within the past 200 000 years, then all the earlier types in Asia and Europe must have been replaced by a second wave out of Africa. This replacement scenario has come to be known by various names, including the 'African replacement' model. Among its many champions, Chris Stringer of the British Museum is particularly well known. Stephen Jay Gould also likes the idea of a sudden replacement. In its extreme form the African replacement hypothesis has the obvious merit of being neat and tidy: all modern humans came from the same African ancestors not very long ago, all earlier types were completely replaced, and therefore all regional differences between human populations must have arisen very recently. No loose ends.

The opposing view at the other extreme is that there was no replacement by a second wave, and modern humans everywhere evolved from the *Homo erectus* stock which emerged from Africa shortly after it first appeared there and colonised Asia and Europe. In this view there is 'regional continuity' both within and outside Africa, and the regional differences that we can see may have developed over hundreds of thousands of years.

Milford Wolpoff, a heavyweight anthropologist at the University of Michigan, and Alan Thorne of the Australian National University in Canberra have been solidly behind this explanation for many years. They were less than impressed by the vision of a recent African Eve—even when it seemed as though the African replacement supporters had genetics on their side. Wolpoff, Thorne and others of the regional continuity school put their trust in fossil evidence which they claimed could not possibly be explained by the story based on genetics. In particular, some of the first Australians, dating from perhaps as much as 60 000 years ago, as well as more recent ancestors, had skulls very similar to those of *Homo erectus* in Java. The brow ridges and heavy bone structure are distinctive and could not have been derived from any types that the African replacement theorists could point to. Thorne and Wolpoff's interpretation is that the skulls are from people who evolved in the Java region from original *H. erectus* stock, just as more gracile types evolved from *H. erectus* in China. Fossils of both kinds have been found in Australia.[16]

The dispute is unresolved, and genetic studies favouring both sides of the debate have now been reported.[17] Both theories could quite possibly have some truth in them. Perhaps some populations were replaced or assimilated while at the same time others were more regionally isolated and retained their identity. Wolpoff can accept this possibility, and other experts think it the most likely pattern of events.[18] I can only comment that the facts of life usually do turn out to be a lot less simple than we first think.

ONE SPECIES LEFT

Whatever the history of regional differences, there has not been enough time, or adequate separation, for any ethnic group to get genetically very far from the parent stock. All living humans are still part of a common gene pool and the genetic similarities between any two races far outweigh the genetic differences. Geneticists have calculated that if all humans except one group, be it the people of New Guinea or New England, were annihilated, we would still be left with 85 per cent of all human genetic variation.[19] And yet zoologists and taxonomists, working with other animals that are geographically varied, readily name subspecies on the basis of smaller and fewer observable differences than can be seen, say, between the native peoples of Greenland and southern Africa, or northern Europe and Australia. The word 'subspecies' is simply the formal zoological term for what in everyday language we may call a race.

In separated populations, as we saw with the tiger snakes, mutations can accumulate independently to the point where the populations would no longer be inter-fertile if they ever came together again. But there is no firm rule about how long it takes. Clearly, it will be faster in real time for those species with short life cycles, even though humans do have a high mutation rate.[20] Bacteria can get through 4000 generations in a couple of months or so but humans need 100 000 years. Speciation, resulting from ecological separation, is also likely to be faster in animals that have very particular needs such as specialised diets or breeding sites. Generalists, like humans, even without much technology, can more easily adapt to new circumstances and opportunities while keeping the same genetic make-up.

In all cases, separate populations are more likely to become different species if there is minimal cross-breeding. Males may kill or avoid each other at the territorial borders, but if a few females have the enemies' babies the rival groups are less likely to become genetically separated. Mixed genes from the margins can gradually spread through the populations and work against any natural selection which is favouring different local adaptations.[21] And as there is evidence that *Homo erectus* was capable of making sea crossings by raft or boat as long as 800 000 years ago, it must be a long time since there were any geographical barriers effective against human gene flow.[22] There are certainly none today.

It is a general principle of ecology (Gause's principle, or competitive exclusion) that where two competing species live together in a stable environment there must be some degree of ecological separation or else one species will eventually eliminate or exclude the other. In view of this there is a need to explain how two species of human could possibly have existed at the same time and in the same place as recently as 30 000 years ago.

NEANDERTHALS AND US

In 1856, just three years before Darwin's *Origin* was published, workers in a limestone quarry in Germany dug up some old bones and tossed them aside. The quarry was in a valley near Düsseldorf known locally as Neander. The old German word for valley is *Thal*.[23] The bones were eventually salvaged and described by German academics but nobody could explain the appearance of Neanderthal man. The anthropologist who described his bones recalled that Latin writers had referred to primitive, wild races from the north. Others, determined to forestall the evolutionists, explained the ancient brow ridges and sturdy bowed legs in terms of mental disorder and physical deformity. One academic declared the remains to be those of a Mongolian Cossack who had deserted from the Russian army during the pursuit of Napoleon. After the next Neanderthal discovery, 30 years later in Belgium, most people were content to believe that there had once been an inferior race in Europe, but they had been replaced by the ancestors of modern Europeans from Asia.[24]

Neanderthal people were living all over Europe and western Asia from about 120 000 years ago, and were possibly present for tens of thousands of years before that. In Israel they apparently coexisted with modern humans for some 40 000 years and in Western Europe they were contemporaries of modern humans for the last 10 000 years of Neanderthal existence.[25] They were stocky people with heavy bones, brains as big as ours, prominent noses and flat chins. They made a variety of stone tools in what is known as the Mousterian tradition, buried their dead with some ritual, and kept themselves warm with fires and clothing made of hides and furs, for they lived through long periods of freezing cold.

Much has been written about the Neanderthals, largely because nobody is sure what became of them. Were they deliberately slaughtered by Stone Age invaders from the east, or did they dwindle to extinction, unable to compete for survival against superior technology? We still cannot be sure that they didn't mate with the moderns and become assimilated. In 1997 a team led by Svante Pääbo, a Swedish expert, took a tiny sample from an arm bone of that first Neanderthal skeleton from Germany. They managed to find mitochondrial DNA and piece together a sequence of 379 base pairs from smaller bits. All modern humans proved to be more similar to each other than they were to the Neanderthal, which suggests the Neanderthals did not contribute to modern DNA. But it wasn't much of a sample.[26]

More recently, a skeleton found in Portugal supported the opposite conclusion by showing a combination of Neanderthal and modern features.[27] This finding suggests that Neanderthals are to be remembered not as a separate species, *Homo neanderthalensis*, which went the way of the woolly mammoth, but as a subspecies of archaic human, *Homo sapiens neanderthalensis*. It also implies, as one reporter put it, that as far as Europeans are concerned there is a little Neanderthal in all of us.

CHAPTER 8

WERE PEOPLE EVER WILDLIFE?

C ONSIDERING THE ASSORTMENT OF fossils from the period before anatomically modern humans appeared, it is easy enough to understand the squabbles between those of a lumping and those of a splitting persuasion. But both sides are agreed that the first anatomically modern humans had reasonably tucked-in faces, a cranial capacity of more than 1200 cm³, and that they generally resembled us more than they resembled anything else, even if they did look rather archaic. They were our ancestors and, given a briefing on behaviour, a visit to the hairdresser and a modern outfit to wear, they ought to be able to pass muster in the New York subway or a queue for a London bus.

We had at least the anatomy, the appearance, of contemporary humanity by more than 100 000 years ago. But what if we could go a bit further back in our evolution, say to *Homo erectus*, and find a few individuals who would not be regarded as entirely human? What then? No doubt there would be a clamour for their DNA, and biologists would determine just how human they were. The point I want to make is that if appearance and the genome make the sum of human uniqueness, then we are unique only in the same way that an orang-utan is unique. Or a lion or a giant panda. We really are just another kind of animal, differing from others along common dimensions. In fact, there would be less difference between us and some other species than there is between a giant panda and its closest relatives.

But I suspect that most people think there is more to being human than the purely physical, and in this chapter we will try to see what else might be special and what it was that made us what we are. As far as science can determine, there can be no mind without a body; and bodies, despite quibbles about particular adaptations, are invariably well suited to the environments in which they evolved. We will make that our starting point.

FORCES OF CHANGE

Plate tectonics is not the only influence on climate. It appears that the old version of the 'four ice ages' which we learned in 1950s geography lessons was a lot less than adequate. It was based on Alpine research begun during the nineteenth century, and we now know that the layers to be found in snow-covered mountains are affected by processes other than climate so they can present misleading profiles. The more recent picture, based on chemical analyses of cores extracted from the seabed, is one of constant change associated with variations in the Earth's orbit and angle of tilt. These operate on a time scale of tens of thousands of years, and no fewer than eight glacial–interglacial cycles have occurred during the past 0.8 million years alone.

A major climate shift believed to have been associated with tectonics and possibly glaciation in Antarctica was a relatively abrupt cooling which took place about fourteen million years ago. Another rather sudden shift, associated with glaciation in the northern hemisphere, occurred around 2.4 million years ago—the time of first appearance in the fossil record of the genus *Homo*. Since these events—shorter cycles and seasonal fluctuations notwithstanding—the world has been a cooler and drier place. It has been drier, on average, because there is less evaporation in cooler times. Glaciation lowers the sea levels as well because less of the precipitation returns to the sea; some of it gets locked up in the ice. We are, of course, talking about enormous volumes of ice. At peak glaciations the biggest ice sheets have been five kilometres thick and heavy enough to depress the Earth's crust by more than a kilometre.[1]

So whatever the ancestors of the first hominids may have been, we know that their line had come through much warmer and wetter times. For most of the Cenozoic, the Age of Mammals, Africa was under dense, tropical forest. In contrast, by the time those australopithecines walked through the volcanic ash of Laetoli, the African Rift was largely under grass, with a mosaic of open woodland, scattered trees or thornbush, riverine forest and huge lakes. The same scene can still be found. It is a mosaic on a vast scale and it supports the greatest variety of large mammals to be found anywhere on Earth. Many of these animals—including lions, leopards, hyenas, as well as *Homo erectus*—gradually extended their range. Within their generations they would have experienced no sudden transitions as they moved out of Africa and into Asia and Europe. Scientists may argue about the ecological status of *Homo* at different times, but no scientist can doubt that our genus was once an integral part of that community of wild animals. Uncertainties about the precise niche that *Homo* occupied will always exist because nobody can work out a complete lifestyle from only a skeleton, still less from bits of skeleton—and fossilised ones at that. Still, an enormous amount of work has been done and not only by the fossil fraternity. We can examine the main issues one by one.

ADAPTED MAN

Sometimes animals that appear to be competing for the same resource are not in competition at all. For example, among African grazing animals the buffalo and zebra will take the tallest, toughest grass stalks and the little Thomson's gazelle will then eat the lowest of the exposed shoots, that bigger grazers miss.[2] Ecological separation can be quite subtle, and in the mosaic of habitats offered by the African Rift there is plenty of scope for it. The period around 2.4 million years ago was also when the environment was unstable on a time scale that promotes evolutionary change, and it was a time when many extinctions and first appearances of species seem to have occurred.[3]

During the early stages of hominid evolution there could have been quite an assortment of ape-like types adapting in their various ways to drier and more open country. Over the generations they tended to spend more daylight time on the ground and rather less time in the comparative safety of the trees. In at least some lines of australopithecines, walking upright evidently proved its worth.

As I said earlier, it takes a lot of physical change to make a four-handed tree-dweller into something that can walk upright. All those modifications of the skeleton and associated muscles must have conferred advantages that outweighed any disadvantages throughout the period of evolutionary change. They must have translated, on average, into a better success rate in leaving surviving offspring. In other words, the changes had to be adaptive. What advantages could there have been to exert such selective pressure?

The australopithecines were a mixed lot in that the robust types had powerful jaws and heavy molars, suggestive of a diet of tough vegetation, while the gracile types had more moderate molars but bigger incisors and canines. There is no convincing evidence that they ate much meat.[4] Speculations about diet could be endless but if, as seems likely, they had to cover a lot of ground to get what they needed, then there are distinct advantages to being upright because walking on two legs is more energy efficient than knuckle-walking. Chimps cover about 11 km a day with similar energy consumption to that of a human who covers about 16 km.[5] Nobody yet knows whether the ancestors of the australopithecines were knuckle-walkers; possibly they took the bipedal path directly from the trees. Either way, an increasingly two-legged position would make sense in terms of energy efficiency. It would not make sense to think of the ancestors of australopithecines as ever having been fully adapted for travel on four feet—as are modern baboons.

An upright body also presents a much smaller surface to the sun and, when this is combined with a dark, functionally naked skin covered in sweat glands, the body is equipped externally for very finely tuned temperature control. And we must not forget that those dexterous hands which evolved in the trees (together with excellent binocular vision and hand–eye coordination) are still free to be used for other purposes. The baboons, on the other hand, which use their hands as feet in the full four-footed position, have evolved a compromise which is a highly successful way to move in savanna country but has closed off, or severely limited, their ability to carry anything and to manipulate what they hold.

Some of the advantages of walking on two legs, then, had been exploited for about two million years by a variety of hominids before further changes occurred that resulted in what we now regard as a more human appearance. Even the hominids that were around when the first recognisable stone tools appeared, such as

H. habilis (or *A. habilis*, depending which leader you follow), had brains only about half the size of those of modern humans.

We don't need to revisit the problems of classification to recall that there have been a number of obvious trends through the evolution of *Homo*, such as changes in size, build, shape of skull and reduction of brow ridges. But the two most regular trends have been towards increasing brain size and smaller teeth. Now, teeth, like guts, are indisputably linked to diet, and the teeth of the more recent humans certainly reflect a move away from hard, coarse materials that needed a lot of grinding. Studies of *H. erectus* teeth with a scanning electron microscope show surface scratches similar to those on a pig's teeth, but it isn't known to what extent they indicate a long-term diet or just the more recent meals of an individual's life. No firm conclusions have been accepted.[6] With the question of brain size not even the linkages are beyond dispute, but before looking at the possible benefits it is as well to consider the costs.

CEREBRAL MAN

Energy is the common currency of all life, and brains come expensive. Our brain makes up only about 2 per cent of body weight but uses up nearly 20 per cent of the energy from basic metabolism—that is, the energy produced in our bodies when we sit at rest.[7] Two per cent of body weight doesn't sound much but it's still a lot more brain than most animals possess. A few words of explanation are needed here. Some body parts increase in direct proportion to body size: if one species of monkey weighs twice as much as another, its skeleton will be about twice the weight as well. So will its heart and lungs. But this will not apply to all parts. For example, in animals with a similar lifestyle the eyes tend to be disproportionately bigger in the smaller animals. The same applies to the brain; it tends to increase at only three-quarters the rate of body size across species. To make useful comparisons a mathematical formula must be applied which allows for this in producing an index or quotient. For the brain it is known as the encephalisation quotient, or EQ. An EQ of 1 means that the animal has a brain of the 'expected' size having taken the sliding scale into account. Elephants have an EQ of 1.03—almost exactly on target. Chimps have an EQ of about 2.4, which is similar to that of australopithecines. Early humans until about a million years ago had EQs in the region of 3.5. This increased progressively to the modern human EQ of between 6 and 7.[8,9]

There is another high cost to having a disproportionately big brain and that is the difficulty of giving birth. When an ape is born its brain will be nearly half the adult size. A human baby with a brain at a similar stage of development would not be able to pass through its mother's pelvic girdle. If women had pelvic girdles that permitted such births they wouldn't be able to walk properly. Human babies are therefore born with brains about a quarter of the adult size. This means that human babies spend a relatively long time, after they are born, in a very immature state. For some months after birth the baby's growing brain uses up something like 60 per cent of its energy—which of course the mother must provide and at a time when her activities, in a hunting–gathering community, are seriously constrained. Big brains are not acquired lightly.

How could these metabolically expensive organs possibly translate into more surviving offspring? There must have been advantage in it, so where did the advantage lie? Science writer Colin Tudge used absurdity to good effect when he asked: 'What for example would a codfish do with the brain of Jane Austen?'.[10] Exactly. Somewhere in our evolutionary history, as Robert Foley puts it, there were ecological conditions that 'allowed the benefits of greater intelligence to outweigh the costs. It is these ecological circumstances that are rare'. Foley, an evolutionary biologist at Cambridge, makes the further point that, regardless of whether the selective advantages were ecological or social, 'conditions that allow benefits to exceed costs must ultimately be energetic'.[11]

It used to be generally assumed that selection for brain superiority was what had kept primate evolution on the path that led some lines to humanity; that bigger and better brains had always somehow conferred the overriding advantage. But this was before we learned that there had been primates striding around on two legs with a brain no bigger than that of a chimpanzee. The brain advantage, according to the fossil record, didn't put any group ahead of the field until long after the advantages of a two-legged gait had been followed up. By then we were also making use of tools, and it has sometimes been suggested that this must have been where the brain advantage lay: that it was the pursuit of technology that set us apart and established the benefits of brain power. This is a convenient point to consider the possibility.

HANDY MAN

Some aspects of non-human behaviour are worth considering if only to see what can be done with brains much smaller than our own. It is now common knowledge that chimps in the wild will prepare a twig by stripping the leaves off to make a better tool for fishing termites out of holes. They also use stones for bashing palm nuts to get at the kernels, or for smashing bone to get at the marrow. They can learn these tricks by observing each other, so that cultural differences arise based on local technology. Cultural variations have been described for about 40 different chimp behaviours in the wild.[12] And this is not confined to chimps. A young female Japanese macaque, in a colony of these monkeys being provisioned in the wild, discovered how to separate sand from wheat by throwing a handful of the mixture into water then scooping the floating grain off the surface. Other youngsters quickly learned the trick and eventually mothers were teaching their offspring.[13] Similarly, in Africa, a female vervet monkey acquired the technique of soaking acacia pods in the liquid trapped in natural tree wells. This softened up the pods for eating. During the drought of 1983 the technique was copied by other female vervets.[14]

An orang-utan at Bristol Zoo was once taught how to knock flakes off a cobble and use them to cut string around a food box, but it wasn't until quite recently that any non-human showed itself capable of striking flakes off a cobble held in one hand with a hammer stone wielded by the other. The animal was a bonobo or pygmy chimp called Kanzi, which lives at the Yerkes Primate Center in Atlanta. Kanzi learned within a day how to use stone flakes to cut string and obtain a reward, but it took researchers months to teach him how to make the flakes in a two-handed way. He preferred to smash his rock by throwing it hard on the concrete floor, or against another rock if there was no concrete.[15] The researchers, Sue Savage-Rumbaugh and Nicholas Toth, acknowledged that Kanji was less skilled than *Homo habilis* had been but were nevertheless enthusiastic about the result. Ian

Tattersall was less impressed. He pointed out that Kanzi 'showed no insight into the properties of the materials he was working with'.[16] True, but not all humans are cut out to be stonemasons either.

Apart from the insight that may be involved in making tools, it is not a far cry from bashing a bone with a rock to doing crude butchering with stone flakes. And we now know that stone flakes can be produced and used by a pygmy chimp—even if it does need a lot of tuition. It is fair to say that one or more individuals of *H. habilis*, or whatever it was, found out how to do something with stone that no other species had discovered. Having made that breakthrough, those early humans continued to make stone tools in exactly the same pattern for well over a million years and became more dependent on tools than any non-human has ever been. So in technology we do find a distinction between the earliest humans and other animals, but it is scarcely a great divide. Nor does it seem that technology did anything to widen the gap. The earliest stone tools are unlikely to have been an intellectual challenge to a creature like *H. habilis* with an EQ of around 3.5. And considering that the technology didn't change for tens of thousands of generations it would appear that tool use lagged behind brain development, rather than driving it through selection for better tool makers. Whatever it was that made brains begin to tell in the lives of those early humans, it must have been something that doesn't show up in the fossils—or the stones.

SOCIAL MAN

In the 1960s a popular image was that of 'man the hunter', in which primitive humans gradually became better organised at preying on large wild animals. In this scenario better brains meant better social organisation and more effective hunting, but then modern technology began to throw doubt on the image by suggesting that many of the animal bones found with human fossils showed the marks of non-human predators. Could it be then that early man had got much of his meat by scavenging the kills of other predators? It wasn't an image likely to fire popular imagination. 'Man the scavenger' has no sex appeal whatever. Nor was the macho male image, as a central pillar of human evolution, likely to appeal much to feminists. They were more likely to welcome the 'woman the gatherer' version put forward by the anthropologists Nancy Tanner and Adrienne Zihlman in the 1970s.[17,18] These authors emphasised the role of females in evolution because the mainstay of subsistence was gathered food, and most of the gathering, as well as the care and provisioning of offspring, was almost certainly done by females.

Speculation could fill the rest of this book but it might be worth making a few basic points before moving on. Among mainly vegetarian species a higher intake of meat does make sense in terms of energy because meat is a very high-energy food. It would also enable humans to function with a smaller digestive system (lower energy cost) than can primates that subsist on coarse vegetation. Gorillas live largely on a diet of leaves, and to cope with it they need a huge windbag of a belly. High-energy vegetarian diets in the wild must contain plenty of fruit or nuts and in savannas these items are not plentiful throughout the year. Baboons typically supplement their diet by working the ground for whatever small creatures they can catch by hand. From what we know of early human ecology, and the shapes and sizes of early man, there are sound reasons for associating bigger brains with a higher intake of meat. How, then, was it obtained?

It is a fact that chimpanzees, with smaller EQs than the earliest kinds of *Homo*, do hunt cooperatively for colobus monkeys and other animals that they could not conceivably catch if they hunted singly. Jane Goodall reported this behaviour and the BBC Natural History Unit has filmed it. And a grisly business it is: the chimps share out the meat by tearing the wretched monkeys limb from limb. I think that anyone who claimed that early humans never got together to hunt, when they were surrounded by meat on the hoof, would need to have a very good argument. If there is one I have never heard it. Not only did early humans have bigger brains than chimps, they also had far more potential for using missiles—beginning with stones and sticks. This we know from the anatomy of their shoulders, arms and hands. Modern humans in hunter–gatherer societies have proven themselves adept at killing even the biggest game animals with the simplest of weapons, so we know it can be done. Of course modern humans have modern brains, but chimps also throw missiles so the idea is not beyond them, even though they are not very effective throwers.

What about scavenging? All modern human societies are known to eat meat from animals that have been killed by other species. Dogs have been used for thousands of years to bring down game. If people happened to see a kill in the wild, why wouldn't they drive the predator off and take the meat? I have personally seen it happen and would readily have done the same myself in the circumstances. I have never been near enough to starvation that I could have eaten the meat of a freshly killed human, but that may be a relatively recent cultural attitude. In 1978 a judge in New Guinea, sentencing three men to fifteen months' imprisonment for taking, cooking and eating meat from a dead man, made the point that in remote parts of the country the practice had once been common in preliterate communities. He added, however, that the narrower references of a single primitive unit of society were no longer acceptable to moderate Melanesians. In 1971 another judge had acquitted seven men brought before him on the same charge, on the grounds that their behaviour had been 'normal and reasonable' in view of their 'very primitive background'.[19] There are cannibal stories from most continents but the remarks of an American judge, in sentencing America's first convicted cannibal, Alfred Packer, in 1873, are hard to believe. Packer, a prospector, survived in a snowbound shack by eating his dead companions. The judge said to him: 'You are a low-down depraved son of a bitch. There were only seven Democrats in Hinsdale County, and you ate five of them'.[20]

Cannibalism aside, it seems most likely that early humans would have obtained meat by both hunting and scavenging. As for the rest of the diet, in all modern hunter–gatherer societies the women, usually working cooperatively, are responsible for gathering most of the fruit, vegetables, shellfish, grubs and everything else that can be obtained without physical violence—though sometimes women help with hunting as well. It seems reasonable to assume that all these things happened in the communities of early humans.

Theorists who have emphasised the importance of one activity or another have usually been concerned to identify the behaviour that would place a premium on social and personal relationships. The assumption is that this would have given a selective advantage to brains that could better cope with social complexity, and that this would have driven brain evolution. Is it a reasonable assumption?

Nearly all existing primates are social animals, and the primates as a group have higher EQs than other animals. Better brains may be essential to deal with social complexity because the species living in the more complex social groups also have the most developed neocortex—the 'thinking brain'. Even animals with very little reasoning power, such as poultry, will establish and maintain social 'pecking orders', or dominance hierarchies as they are more formally called. But it is generally agreed that for an individual to engage in complex social interactions, such as forming alliances and trying to predict the actions of another, it must have an awareness of itself as an individual and be able to see itself from another's point of view. This seems to be outside the experience of most animals; they can react to others but their reacting does not involve an awareness of themselves as individuals. Recently in Australia I have watched a sunbird repeatedly flying at its reflection in a car wing-mirror, a magpie-lark doing the same thing against a shiny hubcap, and a peacock, in full display plumage, hurling itself to the point of injury against a glass door. Each of these birds presumably thought it was attacking a rival. The peacock had to be fenced out for its own protection. Even monkeys, when they see their own reflections, do not recognise themselves.

Chimpanzees are not fooled by mirrors. If a chimp, while asleep or anaesthetised, has a dab of paint put on its face in the interests of science, it will be none the wiser when it wakes up. But when it sees its face in a mirror it will try to wipe the paint off. It evidently knows what it is looking at.[21] In their societies and in captive colonies, chimps have demonstrated that they can set out deliberately to deceive one another, or to form liaisons in a power politics sort of way, or simply to cooperate for the common good. Some sceptics think the behaviour can be explained by simple cause–effect associations without the chimps needing to have any insight into each other's state of mind, but most behavioural scientists are satisfied that chimps show evidence of a lot more social awareness than do any non-primates. Take the behaviour of Yeroen, for instance, a chimp on the island colony at Arnhem Zoo in the Netherlands. Yeroen had his hand cut in a fight with Nikkie, another chimp. For nearly a week Yeroen was limping from this minor wound and generally looking pitiable by any human understanding. But he did this only when

Nikkie could see him. When Nikkie was out of sight Yeroen walked normally. Or the mother chimp who put a finger over her infant's mouth when the infant, presumably feeling secure in mother's lap, started barking with childish ferocity at a dominant group member.[22] Sometimes such things might be explained away as responses acquired by learning from past experience and without insight, but incidents of this sort are commonplace among chimps and they appear in novel situations that haven't arisen before.[23]

So where does this get us? We have evidence that relationships based on social insight and self-awareness can exist among primates with much smaller brains than had the first human types. Perhaps there was a premium on social complexity. In the 1970s Richard Alexander of Michigan University argued that increasing skill in social negotiations would have been favoured by natural selection, so there would have been a premium on higher levels of social intelligence.[24] Alexander and some other sociobiologists (more on sociobiology in Chapter 14) maintain that a more scheming, devious mind would be better able to turn complicated relationships to personal advantage, but not everybody likes this emphasis. Certainly it needs a lot more intelligence than a chimpanzee can muster to be able to handle thoughts like: 'Fred thinks I don't know that Mary knows what Sally did with . . .'. We are left trying to judge how important this sort of thing has been in human evolution.

When you think about it, we do make some pretty complicated social analyses in everyday life, and we do it almost as automatically as changing gear. This is true irrespective of the importance of lying and deceit in securing cooperation and achieving our ends. And when I say *our* ends I mean, of course, those of individuals and that basic social unit—the family. It is time to think about that for a moment, from a biological perspective.

FAMILY MAN

In all primate societies the basic social link is that between mother and infant. There are not many other generalisations that can be made because there is so much variation between species, but it is common for related females to form the most stable and enduring part of a social unit. Males leave this matrilineal group when they mature. This is what happens among savanna baboons, where mothers, daughters and sisters usually stay together for life. Chimpanzees behave quite differently, in that related males make up the core of the group. Females with young often forage independently within the same range as the males, and when they are sexually receptive they may mate with one or several of them. Maturing female chimps are likely to move off and join another group. Gorillas are different again. In this case one dominant male accompanies a group of females and together they form a stable family unit for many years. Maturing males may stay with the group so that the oldest son may eventually inherit group dominance.[25] The risk from predators and the distances that have to be covered to collect food every day are just two factors that will have influenced the particular pattern that evolved in each case.

Among modern humans, cultural differences have long ago replaced the early ancestral arrangement and, as might be expected, there is endless argument about what form it might have taken. There are a few biological clues. As a rule, in those primate species which are monogamous (there aren't many) there is no difference in size and strength between male and female. On the other hand, in those species in which a dominant male has exclusive access to several females, and must defend this position against rivals, the male is usually bigger and stronger than the females. This sexual dimorphism is conspicuous in gorillas— where the male is nearly twice as heavy, and much more powerful, than his mates. In humans, women average about 84 per cent the size of men and are less powerfully built. This could suggest an ancestral pattern in which a single male would accompany and defend several females, although there are other possible explanations for sexual dimorphism.

By one count at least 80 per cent of human societies are at least mildly polygynous, but most marriages are monogamous even where polygyny is acceptable.[26] Clearly, a system that leaves many young men unable to find mates, because the more affluent or influential men are maintaining several wives, is not conducive to peace and social harmony. It isn't surprising that only the rich and powerful in polygynous societies actually live that way.

Not everybody is convinced that polygyny was the norm even among our early ancestors, but support for the view has recently come from research at the University of Chicago. It has long been known that when women live closely together, say as room mates in college, their menstrual cycles tend to synchronise. It has now been demonstrated experimentally that women produce and release odourless chemical signals called pheromones. The pheromones are produced from glands in the skin and there are receptors for them in a tiny organ inside the nose, near the base of the nasal septum. Researchers Kathleen Stern and Martha McClintock took swabs from the armpits of women in a donor group and wiped them under the noses of women in the recipient group. The swabs had been treated with alcohol. In a control group swabs were used which had simply been treated with alcohol and nothing else. Nobody could smell anything except the alcohol, so the experimental subjects all believed they were in the control group. In any case, none of the 29 women in the experiment knew about its hypothesis; they were told that the main purpose of the study was to find a new way of detecting ovulation. The experiment showed that human pheromones produced by one individual affect the length of the menstrual cycle in another individual. The details of the lengthenings and shortenings that occurred support an explanation for menstrual synchrony.[27]

Pheromones are produced by many animals and influence the physiology and behaviour of receptive individuals through the brain and endocrine systems. They play a critical role in the lives of insects but are not to be dismissed among mammals. Hamsters, for example, recognise group members through pheromones, and in mice the level of stress experienced by an individual in a new environment is affected by the emotional state of the previous occupant. We still don't know the extent to which pheromones are involved in the everyday lives of humans, but the Chicago study has proven that at least one mechanism remains and is functioning. The presence of both emitting and responding mechanisms for pheromones is difficult to explain as anything other than an adaptation. It enables sexually mature females to communicate unconsciously, using chemicals, and this suggests that our ancestors did not breed as solitary females with a mate. If the pheromone system evolved among females breeding as a group then there must have been some reproductive advantage in synchronising the group's fertility. This, together with the sexual dimorphism, does suggest that females lived together in groups, defended by one or more males, perhaps in a gorilla type of family arrangement. No great surprises there.

The question still stands: What were the advantages of bigger brains that made them uniquely cost effective to human beings? There is no evidence for any kind of Stone Age arms race that would have made a difference to the survival of our early ancestors, so it wasn't technology. As for social, cultural and family life, examples can be found among non-humans of the sort that may have existed among the earliest humans, so we find only a difference in complexity rather than anything unique. It is this sort of analysis that leads some authors to conclude that there is only one key to uniquely human relationships. That key is human language.

SYMBOLS AND LANGUAGE

We will probably never know when or how language evolved, because it can only be inferred from artefacts that have survived and from what is thought to be the minimal anatomical equipment that shows up in the fossils. Language and symbolism were unquestionably well developed 30 000 years ago among the Ice Age cave painters of Europe—the Cro-Magnons as they are sometimes called, after a fossil site in France—and there are good reasons for thinking that human language began a very long time before that.

The Stone Age artists painted things that mattered to them, and the paintings were no idle graffiti: this was art by any definition. It went on for generations in a 20 000-year tradition, reaching its full flowering in Europe about 17 000 years ago in the Upper Palaeolithic. In the famous Lascaux cave in France the painters, taking their children with them, descended into echoing galleries hundreds of metres below ground, lighting their way with burning torches and flickering lamps made from hollowed stones filled with fat. They carried pigments in skulls and they constructed ladders or scaffolding so they could paint beyond their natural reach. At the Altimira cave, in Spain, the artists took advantage of natural bulges in the rock face to give rounded bodies to the bison they depicted. The meaning and messages in this cave art are a mystery to us, but there can be no question that the work was taken very seriously by those involved.

Images of large mammals predominate. As well as bison there are horses, mammoths, deer, ibex, bears, lions and rhinoceros.

Some animals are shown with spears or arrows sticking out of them and some are heavily pregnant. Humans appear infrequently in the pictures. Some strange semi-human figures might be hunters disguised for stalking or they might represent something more magical. A popular view for a long time was that the cave paintings represented hunting and fertility magic, but detailed analyses have not confirmed this or any other view.[28] There is some order and repetition in the paintings, and it has been pointed out that the proportion of dangerous animals appearing in them, such as lions, bears, mammoths and rhinoceros, decreased over the long period of Palaeolithic cave art.[29] Presumably they had become less relevant to human existence.

We can only wonder how those Ice Age hunters saw themselves in relation to the animals about them. Did they regard themselves as something quite separate from the beasts they feared or preyed on? Or had they a sense of being a different animal as the bear was different from the horse or the mammoth? All that can be said with certainty is that the paintings were special and symbolic for the communities of the time. The people of the Old Stone Age were obviously thinking—and talking—about things beyond the purely functional. And in this they differed from any other form of life on Earth.

HUMAN WILDLIFE?

For more than a million years the human genus lived only among the animals of the African Rift, gathering, hunting and scavenging a varied diet, and following instincts and traditions brought from even earlier times. They lived in related groups centred on mothers and infants and they possibly joined in bigger bands, and met, and almost certainly fought, other groups on the borders of their range. We will never know the sound of their voices or how well they could share their thoughts. They lived and died by the same rules that governed the lives of all large mammals, for they were part of the same ecology. No doubt they would have found some medicinal plants, as have today's wild apes, and they obviously came to depend on simple tools where other animals used teeth and claws. But still they lived in their place, their niche, among the other creatures. They would have taken advantage of fire, as do the hunting birds which swoop and dive for prey along the bushfire margins, but there is no reason to think they could control it. That came much later, as did the improvements in tools, the carvings, the ritual burials, and eventually the painting and the writing. Somewhere, in this distinctly human progress, was the influence of language.

The ability to tell and to learn about things not present must have been of inestimable benefit in the lives of humans, but it is impossible to say when language became a force for change in evolution. The 'creative explosion' in Europe's Old Stone Age did not, as far as we know, coincide with equivalent creativity in Africa or Asia, though it is assumed that the capabilities were occurring elsewhere and might have arisen earlier in other parts of the world. Whatever the distribution, people then were like people now: they had language and art and cooking and bows for hunting. They were limited only by the technology of their time and were human as we know humans to be.

Is it only language then, the use of symbols, which separates us from the rest? If so, what about those archaic folk who might just have got by in the subway or the bus queue? Would they, after all, be installed in a zoo? Or would everything depend on the shape of the larynx? Is language development the measure by which humanity stands or falls? Should the ultimate arbiter be a speech expert?

We shall return to language and brains later, for I don't want to get ahead of our story. It is time now to consider how the people of the late Stone Age brought about the biggest change of all—a fundamental shift in their own ecology.

CHAPTER 9

ALTERING THE LANDSCAPE

LIKE A MIDGE FLYING around a light bulb, the Earth orbits the sun and intercepts only a trifling fraction of the light. Yet if we could collect and store the sun's energy that lands on our planet after its eight-minute journey through space, we could catch enough in less than an hour to meet the world's needs for a year. At any point at right angles to the sun's rays, and just outside the atmosphere, the sun is delivering about 20 000 calories per square metre per minute. This is called the solar constant. The average value all around the Earth works out at a quarter of the solar constant.

The sun's radiation actually includes wavelengths ranging from X-rays (shortest) to radio-waves (longest), with visible light in between. What we see as sunshine is just the visible spectrum, the colours of the rainbow from violet through green to red. The wavelengths of ultraviolet and infrared are respectively too short and too long for us to see, so they are off the visible scale. I say for *us* to see because some animals can make sense of the UV and the infrared as well.

On its way through our atmosphere, much of the UV is absorbed, especially by the ozone layer, and a high proportion of the total radiation is blocked or scattered by cloud and dust. Even so, a high proportion can get through to ground level. In the American south-west, for instance, it can exceed seven million calories per square metre per day during summer.[1] An active human being weighing 65 kg needs no more than about three million calories a day (3000 kilocalories). If only we had a way of collecting this energy we could get all that our bodies need for an active life just by sunbathing for part of the day, but no animal has found a way to stay alive by using energy direct from the sun.

And so we depend on the chemistry of plants. Generally speaking, plants look green because their chlorophyll pigments have absorbed the blue and red ends of the spectrum and are reflecting only the green wavelengths. That might sound as though plants are using a high proportion of the energy that reaches them, but plants, for all their inimitable chemistry, are not very efficient. Less than half the sun's rays can be used for photosynthesis and only around 1 per cent of the energy absorbed is actually converted into plant tissue, though it can be as high as 3–4.5 per cent with some plants in certain circumstances.[2]

Energy can never be converted into matter without some loss, and in nature the conversions are not very efficient. When an animal eats a plant, less than 20 per cent of the energy swallowed as fruit or vegetables is likely to end up as animal tissue, and 80 per cent will be lost as heat or used up in the process of staying alive. At the next step in a food chain, when a carnivore eats a herbivore, the energy fixed by the carnivore will probably be only about 5 per cent. The exact amounts vary with different kinds of plants and animals and even with the individual metabolism of the animal involved. On a personal note, because of my scrawny build and persistent appetite, my wife has been moved to remark that if I were a horse I would probably be taken away and shot.

As a rough estimate, ecologists agree that on average 90 per cent of fixed energy will be lost or used up at each step in a food chain, so that only about 10 per cent is fixed at the next level: 10 000 calories of cabbage would therefore convert into 1000 calories of caterpillar, 100 calories of bird and only 10 calories of fox, cat or human. Looked at another way, with 90 per cent energy loss at each step in the food chain, it takes many tonnes of vegetation to produce one tonne of meat. In nature, food chains don't usually exceed three or four steps, though there may be other reasons for this besides the obvious limits to the energy flow.[3] It follows that carnivores such as cats and hawks can survive only in lower densities than their prey. They can be visualised as skimming energy off the top of an energy pyramid which has plants at the base and prey species in the middle. Human beings eat a mixed diet and take food from every part of the energy pyramid, from leaves and stems to the flesh of other meat-eaters. And not just on land; with simple technology we can take a similar range of food from the water.

ECOLOGICAL MAN

As well as sunshine and warmth, a patch of land must have water and the necessary minerals in order to be productive. All these come together in tropical forests and swamps, which can average well over two kilograms of plant produce per square metre per year, though in forests the edible parts are mainly at the tops of trees. Deserts and freezing steppes at the other extreme may support only a sparse cover of plants, producing less than 100 grams of material a year from a square metre.[4] Big herbivores, like the antelopes and deer which live in unproductive territories, need to cover a lot of ground to collect enough calories. Predators rely on herbivores to collect and process the vegetation but the predators might be obliged to follow the herds through the seasons. Humans, as omnivores, have more flexibility. They may also have to move long distances with the seasons but they can live without red meat indefinitely—provided they can get enough protein from natural stores such as seeds and nuts and from resident small animals like rodents or grubs.

Early *Homo* species managed to survive in a wide range of habitats with only the crudest of stone tools and some control of fire. This level of technology was all that distinguished human ecology from that of other large omnivores. Clearly, some human populations lived in highly productive country and others in less fertile areas. No doubt territorial disputes occurred and were settled, as in other species, by fighting along territorial borders. Skeletons carry the scars. But in separate areas people lived in quite different habitats, and like any other animal they evolved ways of living that reflected local ecological opportunities and constraints. In warm, wet forests or by tropical lakes it would have been possible for people to occupy a smaller area than that needed by similar numbers in less productive country.

In general, families or bands could survive either by moving about their territory each day and camping in a different spot each night so as to spread the foraging impact—as do modern apes—or they could set up camp in one place and return there each night after working over the surrounding country during the day. Nomads could maintain their lifestyle indefinitely, but those following the base-camp strategy would have to abandon a camp and move on when food supplies within a day's reach became depleted. The site might take a year or so to recover. The particular pattern of existence would have been a mix of strategies, each case being determined by the seasonal availability of food and by the number of people in the group. For all that we may talk in descriptive terms, ecology, like evolution, is very much a numbers game.

There are scarcely any people today who live exclusively by hunting and gathering and probably none who survive in that way with only Stone Age technology. Studies that have been done in the past support the generalisations made above. In the forests of Amazonia, aboriginal Indians are thought to have maintained population densities of around one person to five square kilometres.[5] In the thornbush country of Africa's Kalahari Desert during the 1960s, bands of G/Wi bushmen lived at densities of about one person to eleven square kilometres.[6] They seldom stayed in one place for more than three weeks, by which time food supplies would be exhausted within an 8-kilometre radius of camp. A small group of Australian Aborigines, in a relatively productive riverine area of the Northern Territory, lived at a density of one person to 13–18 km^2, but in the vast sandhills of central Australia the population density of a different Aboriginal tribe was only about one person to 90 km^2.[7]

Not only must hunter–gatherers live within the natural productivity of their territory (the carrying capacity), they must also use the resources comprehensively and switch to less palatable foods when the local favourites are finished or out of season. Meat can be a highly seasonal commodity, impossible to obtain at times but plentiful when prey animals are forced to gather for scarce greenery or water. In arid parts of Australia the Aboriginal diet is likely to have been 70–80 per cent vegetable, but at one dry season camp 70 per cent of the daily fare was kangaroo meat.[8]

Regardless of the particular feeding strategies and social arrangements, the fact remains that for 99 per cent of human existence we have been foraging or hunting for our food and accepting the diet that came naturally from the home range that we occupied. This is not to say that there were no interventions at all. In drier regions, where natural bushfires were a common event, there was probably an early understanding of the ecology of fire. Deliberate burning has been used for millennia to bring about ecological changes favourable to humans—such as promoting green regrowth for herbivores. There may also have been simple agricultural practices, such as leaving bits of tuber in the ground to sprout again.[9] But such evidence as exists from modern hunter–gatherers indicates that we survived by much the same foraging principles as do other animals—that is, by eating what we liked best for as long as the good diet lasted, then surviving the hard times by progressively lowering our standards until we ended up eating whatever would get us through to the next good season.[10] Even among recent hunter–gatherers, hungry people do not leave the best things in the wild for others to find.

For all but 1 per cent of our time on Earth, then, we have been an integral part of the so-called 'balance of nature'. It is a much-misunderstood phrase because in reality there is rarely a constant balance. Rather, there is a longer-term position that exists somewhere between the ups and downs of good and bad seasons—a rough equilibrium that zigzags with the floods and droughts, gluts and famines, and build-ups and crashes of local populations. This has been the way of our survival as a species for nearly all our existence. This has been our evolutionary environment; which is to say that whatever evolutionary adaptations we may have acquired as human beings, we inherited them because in the long term they proved to be an advantage for a hunting and gathering existence. Evolution could not have known what was coming after that.

PEOPLE SHAPING UP

Regardless of regional continuities or African replacements, the basic characteristics of our species—the blood pressures, heart rates, hormone levels, sperm counts, monthly cycles, gestation periods, everything from the brain sizes to bowel movements characteristic of humans everywhere—must have been largely fixed by natural selection before regional fine-tuning began to select for local adaptations. We didn't start off with lanky desert nomads, stocky figures adapted to life in the Arctic or people who feel comfortable at high altitudes. But with few exceptions regional adaptations became fixed, because in each region they further contributed to survival in a hunting and gathering ecology.

By the end of the Pleistocene, following the retreat of the last glacial some 10 000 years ago, almost all habitable parts of the world were occupied by people adapted to their region and living in population densities that reflected the long-term carrying capacity of the land. Best estimates indicate a total world population of no more than five million people.[11] The birth rate would have been as high as it could be given the constraints of slow maturity and breast-feeding for long periods (four years for the San nomads of Africa's Namib Desert[12]). But life expectancy was terribly short by today's standards and infant mortality, even without infanticide, would have been very high. That is how the 'balance of nature' works; that is why there has been such scope for natural selection to operate.

Nobody knows exactly why hunters and gatherers took to farming. We can be fairly sure of the circumstances at the time, but sorting out cause and effect is another matter. It seems that the earliest agriculturists were not better nourished than their own ancestors had been. Evidence of protein and vitamin deficiencies in skeletons indicates quite the opposite trend.[13] Nor does it appear that early agriculture offered an easier way of life. Kalahari bushmen, with their population density attuned to the resources, needed to work for less than four hours to collect their daily energy needs from wild mongongo nuts. The Batak of the Philippines could do even better by digging wild yams. Anbarra Aborigines, of northern Australia, with a staple of wild cycad kernels, could collect the necessary calories in two hours.[14] In contrast, imagine the drudgery of cultivating food plants with stone implements and no agricultural traditions, among a profusion of weeds and pests. One could almost believe that the folk of the Neolithic would see more attraction in taking up bog-jogging. The present balance of opinion is that the first farmers were driven by necessity, not choice, but it still isn't easy to explain why it became necessary.

Bad seasons and lean times would be nothing new, but perhaps if the outlook was perpetually bleak and if a few cultivation tricks had already been learned, such as replanting those bits of tuber or sowing favoured seeds in convenient places, then activities of that sort would become more important and new techniques would be learned rapidly because the brainpower was already there. As we saw in Chapter 8, even monkeys can soon catch onto a new technique once the benefits are apparent. If

Neolithic hunter–gatherers were going through continued hard times it may well have been a result of major climatic shifts following the last glacial period. The timing would be right and it may have caused some redistribution of populations.[15] If numbers were pushed beyond carrying capacity in some areas it could explain how people who had the ability to change would be forced to adopt a more settled existence.

Several writers make the point that a sedentary life had to come before agriculture. But not everybody could have stayed in one place, even for a season, until agriculture got going properly. While some individuals stayed at home to tend and guard the crops, others must have continued to search, ever more widely, for wild foods in the overexploited surroundings. Tens of millions of people still live that way today, and we know that hunting and gathering persisted for thousands of years alongside agriculture. It is recorded that in the stomach of an Iron Age man, preserved in a Danish peat bog, there were fragments of 66 different species of plant and only seven of them were from cultivated types.[16] Established ways were not suddenly swept away even in the heartlands of the new life, nor did the transitions always follow the same pattern.

WORKING THROUGH THE STONE AGE

The terms Palaeolithic, Mesolithic and Neolithic (Old, Middle and New Stone Ages) refer to stages of Stone Age culture, but people in different places were not all at the same stage at the same time. The Palaeolithic started with the first recognisable stone tools in Africa and first ended east of the Mediterranean only about 12 000 years ago. The hand-held choppers and flakes of the Oldowan tradition persisted for well over a million years until they were eventually replaced with a wider variety of shapes that could be mounted on wooden or bone handles. By the end of the Palaeolithic there were many traditions in both the Old and New Worlds turning out much smaller, thinner cutting tools shaped for particular purposes—knives, spear points, scrapers and so on. Agriculture was unknown throughout the entire period of more than two million years.

The Mesolithic bridges the gap, technologically, between the Palaeolithic and agriculture. It is associated exclusively with *Homo sapiens* and continues the trend towards smaller, sharper blades. It was a stage characterised in particular by tiny blades called microliths, which were used as barbs and spear tips but could also be set in rows to make longer tools like sickles for harvesting wild cereals. Microliths were in everyday use throughout the Old World by the time agriculture appeared, though some microliths were being made as long ago as 40 000 years ago in southern Africa.

The last phase of the Stone Age, the Neolithic, is usually linked in our minds to the coming of agriculture and to pottery. However, there was a phase before either of these innovations appeared which nonetheless shows some technological advances over the typical Mesolithic. A much-studied group of people known as the Natufians (from a site called Wadi en-Natuf in present-day Israel) were grinding and pounding wild cereals with a variety of stone and bone implements during this period, and were building circular or oval dwellings with dug-out floors and mud walls.[17] Agriculture, as the deliberate domestication of plants and animals, appeared over a period of about 5000 years at centres east of the Mediterranean, and in China, South-East Asia, Central America, South America and the east of North America. The earliest records come from the so-called 'fertile crescent' which curves from the Jordan Valley, north through Syria and southern Turkey, and south between the Tigris and Euphrates rivers in what is now Iraq.

Not surprisingly, the first species to be cultivated and domesticated were chosen from plants and animals that were locally available. Wheat and barley were cultivated at Jericho and other parts of the fertile crescent at least 10 000 years ago, followed by many other plants including flax, peas, lentils and broad beans. Sheep and goats may have been domesticated even before cereals were grown. Domestic cattle show up rather later, but accounts differ at this level of detail.

In China, cultivation began only a little later than in the fertile crescent, with very early records for millet, soybeans, mulberry, pigs and silkworms. The New World took to agriculture much later with maize (corn), beans, squashes, gourds and potatoes. The Americas were short of suitable animals for domestication, but llamas (camel family), guinea pigs and turkeys were early domesticates.[18] Horses were tamed in the Ukraine, and water buffalo in

South-East Asia, at least 6000 years ago. Camels, of one- and two-humped kinds, joined the domestic ranks more than 1000 years after that. The exact chronology of events is inevitably subject to change as new discoveries are made.

I can't leave this section without mentioning dogs, for they are in a category of their own. They have occupied a special niche in human families for a very long time and were sometimes buried in the same grave as a Mesolithic person. But dogs were domesticated long before the Mesolithic and, if recent genetic studies are any indication, they have been our companions since before we started painting on cave walls—which leaves me wondering, rather sceptically, why dogs don't appear in the paintings.[19]

THE FARMING LIFE

The basic units of survival among the first settlers were groups of one or more extended families. Society began to stratify with the coming of a settled existence, and sharper divisions of labour were established between men and women. There was probably less flexibility in this social structure than there had been among hunter–gatherers, but the two lifestyles were not entirely separated because they were still linked by trade. Some quite long trading routes must have been maintained because materials for artefacts, including valued stone and shells, have been found at settlements as much as 1000 km from the nearest sources. Settlements of the time were tiny: a huddle of low dwellings and a few tended plots covered no more than 2–3 hectares at most. Some were as small as 100 square metres.[20]

The earliest evidence of domestication comes from northern Iraq. Bones of domestic sheep, associated with what are thought to have been semi-permanent camps, could be more than 10 000 years old. Permanent agricultural settlements, possibly as old as 9600 years, have been excavated in northern Syria on the banks of the Euphrates near the modern town of Aleppo. At one site of about twelve hectares, known as Tell Abu Hureyra, rectangular mud-brick houses had plastered floors and were divided into rooms. They are thought to have housed single families but were linked by narrow lanes and courtyards.[21] In Jordan a slightly later settlement, called 'Ain Ghazal, grew to be the largest known Neolithic site and housed around 2000 people. The plaster figurines and burial practices of these people suggest ritualised religion and wider social divisions. Some individuals were buried elaborately and were later exhumed to have plaster faces moulded over their skulls. At the other extreme, the skeletons of youngsters and about a third of the adults were found in rubbish pits.[22] By 7500 years ago the residents of 'Ain Ghazal were using pottery for cooking and storing food but the surrounding land was already becoming exhausted. After a few more centuries the site was no more than a watering place for nomads.

There were first-time farmers all over central Europe by 7000 years ago but it took another thousand years for farming to become established along the west European coasts, and even longer for Scandinavians to adopt the new way of life. Various ethnic groups from the period can be distinguished and the evidence suggests to some authorities that the rapid changes in central Europe were a result of invasion and conquest by southern agriculturists, whereas the north Europeans were left to adopt the new practices in their own good time.[23]

In a pattern repeated in many parts of the world, fertile places near water were settled, fortified, fought over, farmed to exhaustion and abandoned. Small towns emerged, better methods of storing food were discovered, and hunting forays by men were planned and became logistically organised. Judging by the fortifications and weaponry that appeared, the towns needed to be defended every bit as much as territories had been defended by ancestral peoples and earlier hominids. The town of Jericho, which as a temporary settlement dates back to the Natufians of the Mesolithic, became one of the earliest centres of agriculture. Even then, as a town covering four hectares, it had to be surrounded by a massive wall and a moat.[24] This was a more intensive way of living than had ever been experienced by human beings. No longer was territorial defeat just a matter of yielding ground and moving on. Now, in one small place, a whole livelihood was at stake. Now, for the first time, there were substantial repositories of stored energy and property which had to be defended. Urban conquest and defence had never before been part of our behavioural repertoire; two million years of natural selection had left us finely tuned for hunting and gathering. In evolutionary terms we were totally unprepared for our own cultural developments. The same can be said about much of the cultural change that has occurred since the Neolithic.

CITY-STATES TO NATION-STATES

The early settlers of Mesopotamia ('the land between the rivers') chose sites along the upper reaches of the Tigris and Euphrates or in the uplands of what is now the Iraq–Turkey border. The lower, dry plains further south couldn't be settled while the spread of agriculture was still being determined by rainfall. Irrigation canals not only opened up the fertile alluvial plains, they offered a stability that could never be achieved while crops were watered only when it rained. The first villages to move onto the plains were probably not strikingly different from others in the 'fertile crescent', but over a period of roughly 2000 years, during what archaeologists call the Ubaid Period (5500–3500 BC), they achieved the status of towns, many of which probably housed more than a thousand people in buildings made predominantly from mud and clay bricks. There were no trees or rocks on the plains. The farmers who worked the irrigated fields specialised in improving domestic stock and selectively propagating an increasing range of vegetables and fruit trees as well as cereals. Fishermen continued to harvest the wild.

New technologies appeared as the irrigation systems became more elaborate. Carts with wheels were invented. The wheels were solid wood to begin with, later to be fitted with leather tyres and surely a big improvement on log rollers and dragged sledges. Ploughs were pulled by draught animals. Copper tools appeared among the stone ones. Later, copper was melted and alloyed with tin to make bronze—which is harder than the naturally occurring metals. Wheel-thrown pottery appeared among the moulded pots. Some containers are thought to have been standardised, suggestive of weights and measures. Markings on stone and clay can be interpreted as the first attempts at record-keeping. By rather less than 5000 years ago the towns had become the first dynastic city-states, defended by organised armies against major military campaigns.[25]

Technological advances, as always, had been accompanied by social change. Some authors believe that the management of irrigation led directly to refinements in the social hierarchy but, as with the origins of agriculture, it is difficult to identify causes and consequences when several things change in tandem. The only existing intellectual framework for life was religion, and its influence became pervasive and conspicuous. Temples, which started

out as modest buildings recognisable only to experienced archaeologists, became great edifices, 80 metres long and dominating
the cities on stepped 'pyramids' (ziggurats). They were literally
the houses of gods, for they provided accommodation, food and
teams of servants to make sure that the appropriate deity would
take up residence and so be approachable. The city ruler was
personally in charge of the temple of the city god, so Church and
State were already inseparable. All this was occurring about 1000
kilometres east of the land from which the biblical tradition would
emerge more than 2000 years later to give rise to the great
Semitic religions of Judaism, Christianity and Islam.

In southern Mesopotamia many of the people from surrounding
towns crowded into about two dozen cities, some of which had
populations of over 10 000 people. The old kinship relationships
of earlier times now existed within sharply defined social classes
and occupations, with prescribed roles for men and women. Priests
and priestesses, temple workers, carpenters, potters, weavers, metal-
workers, builders, soldiers—all were obtaining their daily calories
not by gathering plants or catching animals, but as measured
payment for services. In the daily round and understandings of these
people, ecology had been replaced by economy.

The first people to live by royal rule in city-states were the
Sumerians. A people of obscure origin, they spoke a language
quite distinct from the Semitic tongues of the more northerly
lands. It was the Sumerians in this Early Dynastic Period (2900–
2350 BC) who introduced a system of writing known as cuneiform
script. It took the form of wedge-shaped marks pressed into wet
clay with a stylus, and was used over a period of some 3000 years
for several languages. Cuneiform script was used to write the first
laws, and, although earlier pictorial scripts had existed, it is cuneiform, with some 600 phonemes, which marks the beginning of
written history.

It was war that brought an end to the Early Dynastic Period.
Scholars will never be able to separate legend from the historical
facts of this period, but in 2350 BC or thereabouts a warlord by
the name of Sargon, from Agade in central Mesopotamia, began
his rise to power. After a series of conquests he replaced the
Sumerian royalty with his own people and established the Akkad-
ian Empire centred on Agade. Many scholars consider this to have
been the world's first nation-state. It collapsed after less than

two centuries and power returned to a Sumerian dynasty for the remainder of the third millenium.[26] A famine, salinisation in the irrigated plain and ethnic strife are believed to have contributed to its final collapse. The wider region was subsequently ruled and fought over by Babylonians, Assyrians, Persians, Arabs, Mongols, Ottomans and the British. On the subject of modern Iraq nobody needs any comment from me.

URBAN MAN

The story of Mesopotamia has filled volumes of scholarly works. I had to restrict myself to an embarrassingly short treatment but I decided it was better than nothing because, as the earliest and best-known example of our transition to urban living, Mesopotamia does serve to illustrate the sort of changes that occurred everywhere. Developments in other regions have differed in the details and have not followed exactly the same sequence, but around the world—in southern Europe, Egypt, the Indus Valley, China and Central America—broadly similar chains of events have occurred. Scattered Neolithic settlements were replaced by towns and then by cities and states, which were controlled by hereditary rule and the organised power of religion. The smelting of iron ore was a further milestone which, in Europe, was passed during the eighth century BC and brought cheap and more effective implements and weapons within reach of everybody.

Obviously, the requirements of urban living, even in pre-industrial times, called for a lot of energy beyond that needed to fuel human bodies and provide a small fire for warmth and cooking. It has been estimated that hunter–gatherers would not have used more than 2000 kilocalories in addition to bodily needs, giving a daily energy use of about 5000 kilocalories per head. Advanced agriculturists would have needed more than five times as much, allowing only 1000 kilocalories for transport. The energy was still coming from current plant production—from burning wood and from green plants channelled through beasts of burden.[27]

There is no doubt that as agricultural empires came and went some significant parts of the land were reduced to treeless and eroded slopes or saline flats where irrigation water had gradually brought salt to the surface until crops could no longer be grown. But until relatively recent times there was still plenty of good land in the world that had not been converted to agriculture at all. There were still great forests and mountains and vast landmasses like Madagascar, New Zealand and Australia where neither towns nor agriculture had ever been known. It may well be that pre-agricultural man did exterminate some of the bigger animals of these places, as well as those of Europe and North America, and the use of fire certainly changed the natural vegetation of many areas, but without agriculture the living face of the Earth was still more or less a product of natural selection. Hunter–gatherers, in harvesting their edible plants and catching their animals, had probably been cropping no more than 1 per cent of the total produce in their territories.[28] But, with agriculture, people could eat as much as 90 per cent of what they grew, even though the total produce per hectare might be much less than nature could achieve with a rich mix of trees, undergrowth and wildlife. Agriculture, by channelling the sun's energy as far as possible through edible crops and domestic stock, could support human populations in densities that had never before been experienced—perhaps comparable to traditional Chinese agriculture, which we know could support a family of at least five or six to the hectare.[29] Archaeologists estimate that the cities of Mesopotamia came to house people at densities of 200 or more to the hectare which, without high-rise, is very close living indeed.

How have we responded as a species to this lifestyle transformation? The most obvious response was to produce offspring at shorter intervals. Mothers who didn't need to carry their youngsters for miles every day could be more prolific. And they were. When the reproductive physiology of hunters and gatherers came to operate under the new circumstances it resulted in ever-growing populations.

One may wonder whether any new evolutionary adaptations have appeared in response to these changed circumstances. Ten thousand years is only a few hundred human generations, but we might have acquired a few adaptations to the farming life. It surely can't be coincidence that north Europeans and other ethnic groups which have spent longest herding cattle now have few individuals who cannot digest milk as adults. In populations that came late to dairy produce, such as the Thais, Chinese or the Ibos of Nigeria, nearly everybody still loses the enzyme lactase after about the age of four and is subsequently unable to cope with large amounts of milk.[30] Other ethnic differences may not be adaptations at all; why should it be, for instance, that the women of Nigeria are thirteen times more likely to produce twins than are Chinese and Japanese mothers?[31]

This is not to say that evolution has been unimportant for us since Neolithic times. While we were crowding together with our domestic animals and overriding their natural selection by shaping their bloodlines ourselves, nature's numbers game was rapidly fixing mutations in all our fellow travellers—pests, parasites and disease organisms. Our ancestors, scattered thinly among the other animals of the African Rift, must have had their parasites and diseases just like every other animal, and they probably had a few in common with other species. But when humans crowded together without sanitation and with other species included in the domestic arrangements, they unwittingly created an unsurpassed environment for the evolution and spread of disease.

Lowly organisms have spent millions of their generations under selection with us so, by now, some of them are exquisitely adapted to living with humans. The beef tapeworm is a good example. It survives in the human gut and uses cows as its secondary host to carry it from one human to another. We eat the beef and the cow eats grass that humans have fouled. The tapeworm, in one form or another, can resist the digestive systems of both animals.

Many of the most infectious diseases are now thought to have had their origins in the animals that we domesticated: smallpox from cowpox, measles from rinderpest, tuberculosis from bovine TB, human influenza from pigs and ducks.[32] People such as Australian Aborigines, who were still hunter–gatherers when they first encountered European colonists, had never been exposed to these diseases and their populations were devastated by them.[33] Not that disease organisms need domestication in order to pass between species—especially closely related species. A recent and particularly serious event was the infection of humans by a changeable strain of the HIV virus, which crossed from chimpanzees to humans in Africa.[34]

More than anything else in our agricultural past, it was the presence of disease among the settled and unsanitary throngs that prevented the truly explosive rise in population that was to come later. We shall return to this part of the story in Chapter 11.

CHAPTER 10

DARWIN, GOD AND SOCIETY

FOR 6000 YEARS THE growth of towns was supported by agriculture and fuelled by the energy fixed by plants that were still growing or not long dead. With the exception of some local peat-burning, the human endeavours of entire civilisations were fuelled only by the energy of food and the burning of current biological produce—mainly wood or charcoal. The same sources of energy that were used to build Stone Age huts were used in building the Colosseum, the Vatican Palace and Europe's first universities. After the coming of agriculture, our energy dependencies remained unchanged until machines were invented that could be fuelled by energy fixed in the distant past. Coal was formed from plants which, in the main deposits, had lived long before the time of the dinosaurs.

There was not much forest left in Britain at the beginning of the eighteenth century, but there was plenty of coal. There were also rather fewer than six million people, nearly all living in villages or towns of fewer than 100 000 inhabitants. The rate of urbanisation picked up later in the century and by 1801, when the first census was taken, London had become the only city in the world with a population of over a million, and one in every ten people in England and Wales lived in a town of at least 100 000 people. The scene was now changing rapidly, and well before 1900 Britain had become the world's first urban nation, with the majority of the workforce no longer directly involved in agriculture.

Most people in industrial towns of the nineteenth century lived and worked under appalling conditions by present western standards. Crowding, squalor and disease did little to slow down the birth rate but death rates rose—especially for infants. In 1841 the average life expectancy in Manchester and Liverpool worked out at about 26 years, compared with 41 years for England and Wales as a whole. Londoners, on average, survived to the age of about 36. Towns continued to grow as a result of migrations from the countryside.[1] At their country house in Kent, Mr and Mrs Charles Darwin, as good parents with money and space around them, managed to raise seven of their ten children to maturity.

This reference to Darwin brings me, albeit with the subtlety of an industrial clog, to the main purpose of this chapter—which is not to provide demographic facts and figures at all. In this chapter I want to consider some of the changes in thinking, in relation to Darwinism, that were recorded during and after these times of unprecedented social upheaval. So much has been written about the response of the Church to Darwin's work that it is easy to forget the social philosophers who were at odds with established religion long before evolution became an issue. We shall see what some of the best-known of those social philosophers had to say about the nature of the beast that most concerned them— *Homo sapiens*. It may help to set the scene for Darwin's arrival.

SOULS AND SAVAGES

At the core of any social theory, it seems to me, there must be an assumption, stated or otherwise, about what human beings are—about human nature. As I mentioned in Chapter 1, that brilliant Frenchman René Descartes was determined to show that scientific investigation was the way to truth in what passed for biology at the time. But Descartes, whose dualist beliefs were similar to those of Plato, was convinced that uniquely human qualities were beyond the scientific approach because they were properties of a non-material soul.

An English contemporary of Descartes, the philosopher Thomas Hobbes (1588–1679), was more concerned with social issues. Hobbes had witnessed, in the English Civil War, the bloody breakdown of Europe's first genuine political nation. The war had been largely a Puritan revolution against the established Church and State. In Hobbes' view, social stability would only be achieved when people voluntarily entered into a social contract which placed absolute authority in the state as a legal—political community. No lesser community of any sort—religious, political or even family—

must be permitted to challenge the properly constituted sovereign state. Only within such a rock-solid framework of law would people be secure enough to fulfil their lives as individuals. As one scholar has expressed it, what Plato's *Republic* did for the ancient city-state, Hobbes, with his great work *Leviathan*, did for the modern nation-state.[2]

We know that Hobbes took a distinctly material view of humanity in every aspect. He could accept that the laws of nature were God's commands, but for Hobbes there was nothing ethereal about human nature. Nor did Hobbes think that the differences between individuals amounted to anything very significant. On the contrary, it was because people were all much the same in terms of abilities that they were all a constant threat to one another. He knew from the writing of Aristotle that bees, ants and other 'political creatures' had well-organised communities, but he reasoned that because they are irrational 'the agreement of these creatures is natural; that of men, is by covenant only, which is artificial'.[3] On that basis, Hobbes argued that without 'a common power to keep them in awe, and to direct their actions to the common benefit', people would lead the sort of lives that he believed they had lived in primitive times—lives which were, in his famous phrase, 'solitary, poor, nasty, brutish and short'.[4] Once people had understood this, Hobbes thought, they would be persuaded to unite as a commonwealth under one supreme authority in the interests of their own peace and security. The Church, too, would need to be subordinated to the State because to have it otherwise would be to set up a 'ghostly authority against the civil' and have laws that were based on sins.[5]

Even in Hobbes' time there was evidence from explorers and others that hunter–gatherers were neither solitary nor without social organisation, though life before agriculture could fairly be described as brutish, and for most people it would have been short. But Hobbes is not alone in basing a great social construct on his own ideas of human nature. The Frenchman Jean-Jacques Rousseau (1712–1778) admits to having been influenced by Hobbes (though more so by Plato), and at least one analyst thinks he had no equal in making the political community 'the single most attractive vision for modern man'. The same analyst thinks it 'fair to say that as many popular movements of revolt have emanated from Rousseau's revolutionary philosophy of the general will as from Marx's ideas of the revolutionary proletariat a century later'.[6]

Yet Rousseau's vision of human nature is in some ways quite the opposite of Hobbes'. According to Rousseau's *On the Origins of Inequality*, natural man had originally lived in a simple state of harmony and equality with a human morality provided by nature. Rousseau's image of the hunter–gatherer was the 'noble savage' of Dryden's poetry.[7] It was the coming of agriculture, the claiming of private property and the resulting complications in social relationships which destroyed this state of contentment and, in Rousseau's view, Christianity could do nothing to restore it.

There were, of course, those who sought a more scientific approach to understanding human nature, but they had little other than thought and common observations to work with. The English philosopher John Locke (1632–1704) argued that knowledge must be gained through experience and observation and not just through thinking. This empirical approach replaced ideas of certainty with questions about probability, yet Locke confidently asserted that human beings are born with minds like blank sheets, and that *all* knowledge must therefore be learned. Modern psychology has found a lot of evidence to the contrary.

Among those who later developed Locke's empiricism, the Scottish philosopher David Hume (1711–1776) hoped to explain human nature in terms of general laws, which could one day be formulated from many recorded observations. Like Locke, he accepted that there would always be uncertainty and that we could never claim to have absolute truth. This was not a way of thinking that pleased the Church, but it couldn't be helped. As Hume observed:

> There is no method of reasoning more common, and yet none more blameable, than ... to endeavour the refutation of any hypothesis, by a pretence of its dangerous consequences to religion and morality. When any opinion leads to absurdities, it is certainly false; but it is not certain that an opinion is false, because of its dangerous consequence.[8]

Enough of philosophers. My main concern was to show that in Europe some very eminent thinkers were searching for explanations about human nature, and upsetting the Church in the process, long before Darwin came along. In the rest of the chapter I hope to show that some big thinkers since Darwin have claimed to find support in his work for an even broader range of ideas. On the subject of human society Darwin himself had very little to say, and nothing at all in *The Origin*.

DARWIN'S RECEPTION

Social philosophers notwithstanding, most folk wanting answers to deep questions in the mid-nineteenth century would still have turned to the Church. Britain, for all its political and industrial advances at the time, was still largely an agrarian nation, where the parish was the familiar administrative unit and the clergy were relatively learned men with some local influence. Not all the clergy in Victorian times were insisting on a literal interpretation of Genesis: the geologists before Darwin had already convinced the more liberal-minded Christians that the world hadn't been created in seven days. But few would have questioned the central argument of the Natural Theologians that the wonders of life were living testimony to the powers of a Great Designer. David Hume had already refuted the logic of this, but he hadn't been able to come up with an alternative.[9] So the Church was not entirely united in its reaction to the publication of *The Origin*, but it was such a vigorous response that we shouldn't be surprised if it contributed to the book's commercial success.

Darwin's work was not just another intellectual challenge to religion based on abstruse arguments about the nature of truth or reason; it was written in language that everybody could understand and was founded on thousands of carefully marshalled little facts and observations from the living world. And not only did it contradict the biblical account of creation, it was a threat to the Natural Theologians' arguments for the existence of God.

The case for Natural Theology had been set out by William Paley (1743–1805) in his book *Natural Theology*, published in 1802. To Paley, Anglican priest, philosopher and Fellow of a Cambridge college, it seemed self-evident that if you found a watch abandoned on the heath, you would know there had to be a watchmaker. So it was with the natural world; the intricacies of nature proved there was a God and He must have had a purpose in mind, otherwise how could you make sense of nature's wonders?[10] The Victorian passion for natural history laid some useful foundations for biology, but before Darwin the glass-cased displays and insect collections would all have been seen from Paley's point of view. *All Things Bright and Beautiful* was written eleven years before *The Origin* came out.

Most of the public furore (and the attention of cartoonists) was centred on what Darwin had not even mentioned in *The Origin*—the obvious conclusion that humans were cousins to the apes. This was distorted in the popular press and in the minds of some of Darwin's opponents, such that evolution became 'the monkey theory' and modern apes or monkeys became our alleged ancestors instead of our contemporaries. It was in the spirit of this popular misrepresentation that Bishop Wilberforce, at the Oxford meeting of the British Association in 1860, asked T.H. Huxley whether he claimed descent from an ape on his father's or his mother's side. It was Huxley's determined defence of Darwin that earned him the title of 'Darwin's bulldog'. The public fuss eventually died down and, in the main, Christians managed to assimilate evolution just as they had earlier come to accept that the Earth is not the centre of the universe; but in the scientific community Darwin also had his critics, especially among older scientists with entrenched opinions to defend. Younger scientists gave *The Origin* a warmer welcome, and within a decade the theory of evolution was scientifically respectable. As well as younger scientists, Darwin acquired distinguished older champions in Germany and on both sides of the Atlantic.

Darwin once wearily remarked that his mind had become 'a kind of machine for grinding general laws out of large collections of facts'. His was a world of biological details, and he never suggested that his theory should be applied to the shaping of human society. He spoke of 'the struggle for existence' with regard to both plants and animals, but when Herbert Spencer, in 1864, coined the famous phrase 'survival of the fittest', in the sense of 'best fitted', Darwin thought it accurate and convenient and included it in later editions of *The Origin*. It's a pity that he did so, because I suspect it has been the most misused quotation from all Darwin's writings.

After *The Origin* was published, Darwin became engrossed in the adaptations of orchids, and had little to do with the social theorists who were taking what they could use from his book, or attempting to find a link with Darwinism that would give scientific status to their ideas.

DARWIN TO THE RIGHT

When the American edition of *The Origin* was released in 1860 it was widely reviewed, but its launch was overshadowed by events leading up to the American Civil War, which began the following year and lasted until 1865. American scientists, with the notable exception of the Harvard zoologist Louis Agassiz, were generally receptive to Darwin's work, while church congregations had other things to worry about. When evolution eventually began to feature in public thinking Roman Catholics were positioned against the idea, and southern fundamentalists are still holding out against it. But the more liberal of the Protestant denominations soon accepted that evolution was the way God worked. Inevitably, biological principles were misapplied in the social context, but at the time they seemed to sit well enough with the American enthusiasm for commerce. The publisher of the *Christian Union*, Henry Ward Beecher, was happy to observe in 1885 that 'design by wholesale is grander than design by retail'.[11] By the 1880s, American Protestants had not only accommodated evolution, they had given it divine purpose.

It is said that Herbert Spencer (1820–1903), with his sweeping synthesis of evolution, physics and social philosophy, had a lot more influence in America than did Darwin himself, though he was less influential in his native England. Largely self-taught, Spencer trained for a time to be a civil engineer but became instead a subeditor of *The Economist*. He resigned that position after five years when he came into an inheritance that enabled him to concentrate on writing and generally furthering a vision of the world according to Spencer. In the words of one authority: 'In the three decades after the Civil War it was impossible to be active in any field of intellectual work without mastering Spencer'. Spencer had a hatred of state power and a passionate faith in individualism, which found a hearty welcome among Americans of the time: 'Spencer's was a system conceived in and dedicated to an age of steel and steam engines, competition, exploitation, and struggle'.

The scientific basis of Spencer's world view came from his

understanding of the laws of physics and the process of evolution. Scientists before Einstein had stated that energy could neither be created nor destroyed but could be changed repeatedly from one form to another. Spencer's understanding of evolution was Lamarckian and progressive so, as he saw it, energy and matter were constantly being transformed through the working of evolution into superior forms of societies as well as species. In societies, which have no fixed lifespan, the ultimate state of complexity and perfection could be reached and maintained indefinitely. As for religion, that came in the 'unknowable' category, so it wasn't an issue for him. Spencer's take-home message was very clear: don't let the State interfere with nature's progress. Charity was all well and good but state-funded assistance to the poor would only impede nature's approach to perfection. Life was, and should be, an individual free-for-all in which the fittest would survive.[12] John D. Rockefeller saw nothing to quibble with in that.

America's home-grown disciple of Spencer was William Graham Sumner (1840–1910), who became Professor of Political and Social Science at Yale and is said to have been the 'most vigorous and influential social Darwinist in America'. The son of a labourer, Sumner was raised to believe in hard work and frugal living. He must also have had a touching faith in natural justice, because his way to get rid of poverty was to 'Let every man be sober, industrious, prudent, and wise, and bring up his children to be so likewise and poverty will be abolished in a few generations'. In Sumner's view there was no point in trying to change society because the great tide of evolution could not be turned by passing laws for social reform. In Sumner's own words:

> Let it be understood that we cannot go outside of this alternative: liberty, inequality, survival of the fittest; not-liberty, equality, survival of the unfittest. The former carries society forward and favours all its best members; the latter carries society downwards and favours all its worst members.[13]

It is probably fair to say that in social Darwinism people found a philosophy that sat well with existing attitudes rather than an irresistible new faith. In its most extreme form it couldn't possibly last. The modern industrial state is too complex to be managed by the rugged individualism of American traditions. The trouble was that as the social economists moved towards a fairer view of society, those with national and racial attitudes came to prominence, and they also claimed support from Darwinism.

Francis Galton (1822–1911) was Charles Darwin's cousin. Another man with a legacy, Galton was free to indulge a broad range of interests, which included meteorology, the use of fingerprints to catch criminals, and developing statistical techniques. He also found time to explore remote parts of Africa, where, we are told, he studied the shape of women's buttocks using a sextant and the principles of surveying.[14] Galton, who was obviously a bright fellow as well as an eccentric, wrote hundreds of scientific papers and nine books. Two of his books, *Hereditary Genius* (1869) and *Inquiries into Human Faculty* (1883), reflected his interest in the inheritance of intelligence. Having satisfied himself (and convinced his cousin Charles) that exceptional ability was inherited, Galton alerted the western world to the possibility of improving society, physically and mentally, through selective breeding—limiting the birth rates of the least fit and encouraging the best breeding stock to marry young and get on with it. It was an idea which he called 'eugenics' (Greek *eugenes*, 'good stock'; as in the name Eugene, well-born).

Galton's ideas aroused no great hostility at the time. He wasn't trying to breed up an elite race of athletes or superbrains; he was explaining what he thought was a way to improve the whole of society. It probably sounded like common sense to those who were in no doubt about their reproductive worth. More practically, Galton founded the world's first human genetics institute as the Laboratory of National Eugenics at London University. Now called the Galton Laboratory, it is still an important centre for human genetics.

First in England, then in America, the eugenics movement gained strength. In 1914 in the USA, a National Conference on Race Betterment 'showed how thoroughly the eugenic ideal had made its way into the medical profession, the colleges, social work and charitable organizations'. By the following year a dozen states had adopted sterilisation laws.[15] If this was going too far for many Americans it could scarcely have been better ammunition for the sociologists, who had long been critical of biological explanations of society. For them, society was more than a gathering of individuals whose collective worth would stamp society for better or worse. Society had a life of its own and had to be studied, explained and shaped at the social level. Levels of study, it seems, were forever to be kept apart.

If the eugenics movement had survived long enough it would presumably have come to terms with some findings of modern genetics that reveal its limitations. We now know, for example, that many genetic problems show up only when two recessive alleles come together. If the recessive is present in 1 per cent of the population, and it can ·be weeded out only by preventing double-recessive individuals from breeding, there will still be half as many people with a single recessive after 100 generations.[16] Referring to the champions of the eugenics movement, the geneticist Steve Jones remarked: 'They were filled with extraordinary self-assurance. Their views were taken seriously although in retrospect it is obvious that they knew almost nothing'.[17] One can only wonder what the geneticists in 100 years' time will be saying about today's social theorists.

As it happened, the eugenics movement failed to survive long enough to be discouraged by modern science. Two world wars, and the horrendous outcome of the Nazis' belief in Aryan racial superiority, made genetics a dirty word for decades. A generation of American anthropologists, following the lead of German-born Franz Boas in particular, took a culture-centred view of human nature in which it was a virtue to renounce all recognition of biological influence.

DARWIN TO THE LEFT

Having always been more interested in biology than political philosophy, I can remember when the word anarchist would have brought to my mind only a cartoon figure wearing a black cloak and carrying a clearly labelled bomb. But for some anarchists that would be quite the wrong image.

A happy state of anarchism was pursued by a Russian prince, Peter Kropotkin (1842–1921), after he renounced his title and became a revolutionary. The ex-prince escaped from a Russian prison only to spend three years locked up in France. He then settled in England, where he remained a free man and did most of his writing. Violent revolution was never part of Kropotkin's program; he was opposed to Bolshevism and much of Marxism.

He had no liking for centralised power, whether capitalist or communist. At the centre of Kropotkin's vision of anarchism was the principle of cooperation between individuals in their communities. He could see the necessity for some large centres of heavy industry but, as far as possible, settlements should remain small so that a sense of community, based on family units, kinship and mutual aid, could remain strong. British and French anarchists had said similar things a century earlier, but to support his thesis that this was the natural way for mankind, Kropotkin cited Darwin's work.

Kropotkin's book, *Mutual Aid: A Factor in Evolution* (1902), is said to be his greatest. It is an ambitious survey of what Kropotkin interpreted as mutual aid throughout the animal kingdom and in different stages of human society. According to Kropotkin, cooperation, not strife, was the watchword of nature, at least within species. But although Kropotkin thought Darwin had made that clear, it seemed to him that other Darwinians were taking away the wrong message.[18]

Presumably Kropotkin would have considered Karl Marx (1818–1883) to have been among those who misread Darwin's work. Yet Karl Marx was so taken with *The Origin* that he wanted to dedicate to Charles Darwin the English edition of *Das Kapital*. Darwin didn't give permission.[19] It was the 'struggle for existence' which Marx found most appealing. In a letter to Friedrich Engels he wrote: 'Darwin's book is very important and serves me as a basis in natural science for the class struggle in history'.[20] It was this idea that biological survival in nature somehow mirrors the social class struggle in communist ideology that established a role for Darwinism, or rather another distortion of it, in Soviet politics.

At this point, it becomes important to recall that Darwin never knew about Mendel's work on genetics. He was, like nearly everybody else, guessing about the nature of inheritance, and was increasingly inclined to favour some sort of Lamarckian explanation. So, too, was the Soviet Politbureau during the 1930s, long after Mendel's work had become widely known. Mendelism didn't please the communist ideologues. The idea of innate differences was distasteful to them; it was 'idealist' and awkward to accommodate in Marxist philosophy.[21] At that time an agricultural technician called Trofim Lysenko had come to some prominence in the Soviet Union as a result of his work on the seasonal cycles of wheat and other food crops. He had been treating seeds by soaking and chilling them in a process which he called vernalisation, and from the results Lysenko felt able to promise vastly improved summer harvests from spring-sown seeds.[22] The icing on this wondrous cake was that vernalisation was a purely environmental treatment; there was no selective breeding involved so the idealist and reactionary genetics of the West could confidently be condemned and rejected.

Basing his conclusions on pilot trials, without proper statistical controls and analyses, Lysenko went from strength to strength, but he never failed to pay homage to the source of his ideas—a horticulturalist called Ivan Michurin. As a keen amateur, Michurin had earlier convinced Lysenko that the process of heredity in plants could be 'destabilised' so as to make plants quickly responsive to environmental change and able to 'assimilate' such responses into their hereditary pathway. It was, in short, a special form of Lamarckism—a sort of adaptive heredity that didn't involve genes.[23] No doubt Lysenko genuinely believed in this, but it couldn't have fitted better with Soviet ideology: Michurin's pioneering work, recognised and developed by Lysenko, had shown to the world that, with living things, the desired result could quickly be achieved by imposing the correct conditions. Lysenko called it 'Soviet creative Darwinism'.

Michurin died in 1935 at the age of 80. Thanks to Lysenko, his fame continued to grow. Lysenko himself was elected President of the Lenin Academy of Agricultural Sciences in 1938, replacing Academician Vavilov, whose slow, careful research in genetics was respected in the West.[24] Vavilov was later accused of being a British spy and was sent to Siberia, where he died after a year or so. In August 1948, at a meeting of the Soviet Praesidium of the Academy of Science, Lysenko's Michurinism was made 'official science', and the 'reactionary ideologist theories' of western genetics were officially rejected.[25] Lysenko continued to claim that his was the true Darwinism, and he declared in his report to the Academy: 'The appearance of Darwin's teaching, expounded in his book, *The Origin of Species*, marked the beginning of scientific biology'.[26]

The decisions of the Academy were reported in Paris on 26 August 1948 as front-page news in *Les lettres françaises*. The item read: 'A great scientific event: heredity is not governed by mysterious factors'.[27]

There was never a public U-turn in Soviet genetics. Gradually, and at great cost, it became obvious that Lysenko's methods were never going to deliver the goods. When spectacular failures couldn't be explained away any longer, and in the light of Watson and Crick's breakthrough with the DNA molecule in 1953, official policy just had to put different strategies under the same old banners. When Premier Khrushchev was praising 'Lysenko's methods' he was likely to be talking about hybridisation or some other Mendelian techniques which Lysenko was, by then, more than willing to endorse. Lysenko followed Khrushchev out of office in early 1965.[28]

EVOLUTION EVERYWHERE

If nothing else, I think the previous two sections make two important points: first, an awful lot of what gets called 'Darwinism' in the popular media has precious little to do with anything that Charles Darwin ever did or said; and second, we are reminded how widely social philosophies have diverged—how widely we have been casting about for a better society since the dislocation of those early forms of human community, which, as far as we know, were based on kinship and religious attachment to ancestral places. Over the centuries, would-be social engineers have put up one scheme after another as solutions to social problems in societies that have necessarily been moving further and further away from the common ancestral pattern. There has been an underlying assumption, or even a clear assertion, that each scheme has the true measure of human nature, and this has ranged from Hobbesian competitive nastiness to a peace-loving romanticism. For a start, this begs the question (in the original sense of that phrase) that there is such a thing as human nature.

Those who study different human societies like to point out that social differences are essentially cultural and therefore learned. This is true, of course. One or two persistent racial differences have been reported, but almost without exception we all behave according to the culture in which we grow up, even when that is different from the culture our parents experienced. But it's also true that humans everywhere behave differently from other social species. In no human society do people behave more like social antelopes or social rodents. In the words of Edward O. Wilson, a Harvard biologist: 'To adopt . . . the social system of a nonprimate species would be insanity in the literal sense. Personalities would quickly dissolve, relationships disintegrate, and reproduction cease'.[29] Clearly, we have evolved a human nature that is distinctly human at the species level.

We are perfectly justified, then, in trying to analyse just *how much* of our nature is the outcome of our evolution as a species and therefore a part of our genetic make-up. Anthropologists have documented what they call cultural universals—characteristics that have been found in every known culture, such as bodily adornment, cooperative labour, food taboos, funeral rites, faith healing, games, jokes, religious ritual, marriage and sexual restrictions. To some scientists, the fact that all cultures everywhere have certain things in common suggests that we are all expressing, in our different cultures, a common evolutionary inheritance. Well, it may suggest it, but it doesn't prove it. Anthropologists can counter the argument by claiming that cultural universals are ancient inventions which have spread through all groups; or have been acquired independently, many times, as cultural adaptations.[30]

With other species it is possible to do experiments, involving controlled environments, social interventions and selective breeding, designed to reveal genetic influence on what looks like learned behaviour. Zoologists have been doing such work for decades and calling their specialisation 'ethology'. Three ethologists shared the Nobel Prize for Physiology or Medicine in 1973.[31] The evolutionary approach to social and individual behaviour studies in wild animals under natural conditions is now a substantial part of what is often called behavioural ecology or evolutionary ecology. Obviously, zoological research methods cannot be applied to humans, so we might well ask whether there is any value in even thinking about our own species in the same way. Many scientists think there is.

In 1975, the above-mentioned Edward O. Wilson wrote an excellent book on social behaviour in vertebrates. He had already done one on insects. He called the vertebrates book *Sociobiology: The New Synthesis*.[32] It would probably not have attracted much attention outside academic biology had it not been for the fact that he devoted his last chapter to humans, thereby marching boldly into a zone where extreme positions seem to be joined by tripwires rather than middle ground—culture v. genes, free will v. determinism, nature v. nurture, science v. humanity, left v. right. The subsequent uproar spread to every quarter where there was a social, ideological or political emotion to be triggered by Wilson's chapter. I suspect that nobody was more surprised than Professor Wilson, but in terms of fame and fortune it cannot have done him any harm at all. He followed up in 1978 with his book *On Human Nature*, which won a Pulitzer Prize.

The reaction to *Sociobiology* was surprising. After all, as Wilson readily acknowledges, it was not a new idea: earlier authors, including Desmond Morris and Robert Ardrey, had deliberately sensationalised the biology of human nature, but they had been ignored or derided by social science academics. Wilson's work, with its textbook status, had to be taken seriously. It is a curious fact that the social sciences are still not united with the natural sciences and are popular with non-science students. Indeed, those sociologists and anthropologists who avoid integration with the natural sciences often write about human behaviour as if we are not any species of animal at all. But this situation is changing as more social scientists come to accept, as do the authors of one recent textbook on cultural anthropology, that 'any serious inquiry into human behaviour will have to consider the complex ways in which cultural and biological processes intersect'.[33] Presumably these authors, and like-minded academics, would concur with Edward Wilson in predicting a sterile future for any social science discipline that remains isolated from advances in biological research.[34]

The storm about sociobiology has recently been written up and analysed in historical detail by an author who followed the controversy at first hand and is herself a professor of sociology.[35] We shall return to Wilson and his sociobiology in Chapter 14.

CHAPTER 11

EXPLOSIVE TIMES: HUMAN ECOLOGY IN THE TWENTIETH CENTURY

A NIMALS IN THE WILD might survive injury and have periods of sickness, but they can rarely remain incapacitated for long and it must be rarer still for any to become infirm through old age: wild animals do not survive to be well past their best. We cannot be sure exactly when human communities first started helping individuals to stay alive when they could no longer support themselves. A Neanderthal skeleton with a withered arm was found in a grave in Iraq dated at 50 000 years, and some see this as evidence that the individual, an old man, must have been cared for in life as well as having been given a burial.[1] It is also difficult to say exactly what the average life expectancy would have been in prehistoric communities, but hundreds of skeletons in two prehistoric cave cemeteries in North Africa gave almost the same survivorship profiles as were obtained from more than a thousand skeletons of an extinct hunter–gatherer group at a site in North America. And the same profile emerged from a census of Yanomami Indians in southern Venezuela that was conducted in 29 villages during the 1960s, when the inhabitants were still hunter—gatherers and had barely encountered any other way of life. In all these populations, three extinct and one living, the average estimated life expectancy was between 19.9 and 22.4 years.

Experts can, and some do, argue about the age estimates for adult skeletons, but there is not much room for disagreement over the ages of immature individuals. Infant mortality was about one in four except for Yanomami females, which was more than one in three because of infanticide. In all four populations at least another 10 per cent of children died before the age of 4 and, of the survivors, only 75 per cent reached puberty, perhaps a third surviving through their reproductive years. In prehistoric times it would have been exceptional for a newborn baby to have two or more surviving grandparents.[2]

Today, more and more of us are pushing the limits of old age, and a substantial research effort is underway to find out how the maximum human lifespan is fixed. It is incidental to this chapter, but worth mentioning, that research into ageing is just one aspect of the massive research commitment in cell biology. It seems there is a limit to how many times our cells can divide, because the cells in an ageing body are not only accumulating copying errors at each division—they are also counting down to that final replication at the end of the dying program. There is a keen scientific interest in the DNA that forms sequences called telomeres on the ends of chromosomes. Babies have over a thousand of them but we lose ten or 20 with each cell division. When there are very few left, the cell dies as if its sands of time have all run out.[3] Yet programmed cell death helps to shape babies to begin with. If embryonic cells between our fingers didn't die and disappear, for example, we would be born with webbed fingers.

AGES AND NUMBERS

There are lots of stories to illustrate the astonishing results that you get from repeatedly doubling a number. If you could fold a sheet of paper just 20 times you would have a wad about 100 metres thick. If you could fold it 42 times your wad would be too thick to fit between the Earth and the moon.[4] Or consider the effects of doubling in relation to time: if dividing bacteria double every hour, then no matter how long they need to half-cover a surface, they will take only one more hour to cover it. In the biblical account, when the Lord urged people to be fruitful and multiply, the world population stood at two. The first doubling would have brought it to four. The last doubling added three thousand million people to the population in less than 40 years. There are now six billion of us.

The so-called population explosion is mainly a twentieth-century phenomenon resulting from the successes of medical science. There were no more than about 1.5 billion people in the world in the year 1900 and fewer than half a billion during the seventeenth century. Rapid population increase was welcomed by governments and other institutions that saw strength in numbers. In 1917 Margaret Sanger, herself the sixth of eleven children, was imprisoned for a month in the USA for promoting birth control. In England, a decade later, Julian Huxley was reprimanded by the director of the BBC for mentioning the same unspeakable subject on the radio.

Our present numbers were correctly predicted four decades ago as we continued to welcome death control while producing families well above replacement level. Sounding alarms in the early 1960s, Julian Huxley was aware that many people still saw nothing to worry about. The World Health Organization had already been twice prevented by Roman Catholic pressure from even considering population density as a health issue.[5] In 1985 pressure from the same source stopped all US funding for the United Nations Fund for Population Activities and the International Planned Parenthood Federation. American contributions were not restored until 1993.[6]

Growing populations have a built-in momentum; they can continue to grow for decades even after the average family size has fallen *below* replacement level. This is easy to understand if you imagine a population to be a young couple who produce only one child. Although one child will not replace a couple, for as long as the couple remain alive the population is 50 per cent bigger than it was. The population structures of most African and Asian countries are heavily dominated by the younger age categories so they are likely to continue increasing for years to come—even if they could acquire overnight all the circumstances that have led to population stability in a few western nations.

We shall look at some global projections and predictions in Chapter 16. In this one I want to outline the existing situation in terms of human ecology.

FEEDING OURSELVES

For all but the last 1 per cent of human time we have gathered and caught our meals from a global diet that included thousands of different plants and virtually every kind of animal from grubs to big game. Then, after settlement and agriculture, people all over the world came to depend on a couple of hundred plant species and a few species of domestic animals. From this small selection, agriculturists everywhere chose the best genetic material for local conditions. Centuries of traditional agriculture teased out thousands of different strains of these domestic plants and animals until there were some types specifically adapted to every agricultural locality. But, sadly, traditional agriculture could not keep pace with burgeoning populations. If we still had to depend for food on our heritage of traditional agriculture, precious though it is, there would be far fewer than six billion of us alive today.

The transformation, which came to be known as the Green Revolution, was brought about by replacing traditional local strains by a few new ones which could produce much higher yields in response to irrigation, synthetic fertilisers, pesticides and mechanisation. In short, the Green Revolution rested on using technology to override local, natural constraints instead of adapting to them in traditional ways with appropriate genetic stocks. With modern farming, a hectare of land can be made to support up to about 20 people instead of the few that could be fed by traditional methods.[7] This has enabled all nations to continue increasing their populations, although, predictably, some countries have benefited a lot more than others from the Green Revolution.

Because some nations wax fat while others, despite the new agriculture, still face famine, it is sometimes claimed that there is no food problem at all—just a distribution problem. This observation only serves to draw attention to the complexities of international politics, territorial borders, religious and ethnic rivalries and global economics—a tangle of notoriously intractable problems. In any case, even in the perfect and egalitarian world of make-believe, the ecological impact of feeding so many people would still have to be borne, and this chapter is about ecology. Ecological wellbeing and social justice are separate issues; the latter is a purely human construct, whereas the former can be as impersonal as the weather. The twin goals are commonly interrelated but they are not the same thing and don't necessarily hang together, as some writers like to imply. In principle, it is quite possible to move towards one while slipping away from the other; we could build a deplorable society in a garden of plenty just as we could create an ecological wasteland in which to starve equitably.

Coupled with international trade, the Green Revolution not only permitted all countries to become more populous, it also enabled nations to import food so that populations could soar beyond the carrying capacity of their lands. Extreme cases, such as Singapore, Malta and Bermuda, have more than 1000 people to the square kilometre so they couldn't even try to feed themselves, but many countries now need to import the bulk of their food. It comes in shiploads, mostly from the granaries of the northern hemisphere. Only fifteen species of cultivated plants now provide 90 per cent of human food energy, and two-thirds of this comes from just three species of grass: wheat, rice and maize.[8]

Ultimately, when the most productive crops have been given all the water and nutrients they can use, the limit to agricultural production is still set by the efficiency of photosynthesis. The Green Revolution brought us closer to this limit while human populations increased along with the food supply. In 1950 there were 0.23 hectares of grainland for every person in the world; it now works out at half as much. Some countries that were self-sufficient in food in 1950 can no longer survive without imports. In Japan and South Korea, for example, grainland now works out at no more than 300 square metres per person and these countries must import 70 per cent of their grain.[9] Globally, we are still ahead in terms of food production, because each hectare is being made to produce more than twice as much as it did before intensive agriculture. But in poor seasons there isn't much to spare, and further technical improvements have been showing diminishing returns as natural limits are approached.[10]

As a rule of thumb, adopted by the World Health Organization (WHO) and the Food and Agriculture Organization of the UN (FAO), the basic nutritional daily needs of adults can be met by diets providing 2200 kilocalories. On this basis, over 800 million people are chronically undernourished.[11] If we all lived on a mainly vegetarian diet, with the minimum necessary protein coming from intensively farmed and high-yielding animals, notably pigs and poultry, then current production could provide an adequate diet for all six billion of us. But existing food production will not provide an affluent western diet for more than about 2.5 billion people.[12] As things are, roughly two of every five tons of the world's grain is fed to farmed animals, mainly to provide dairy products and red meat for those who can afford them.[13]

Cattle and other big herbivores are not the most efficient animals when it comes to converting their food energy into protein for humans. This is in accordance with the energy losses referred to in Chapter 9. But according to health experts the western diet is not only extravagant, it is unhealthy to the extent that it includes more fat and sugar and a lot less roughage than did the diets for which our systems evolved.[14] Yet the wealthier individuals in poorer countries are westernising their diets with such enthusiasm that 'there is no historical precedent for so many people moving up the food chain so fast'.[15]

There was a time when the oceans were regarded as an inexhaustible supply of protein; after all, the Earth's surface is 71 per cent sea. The only problem was how to catch the fish. Today there are some 3.5 million fishing vessels, and modern technology is used to locate and net the shoals. The total fish catch is five times what it was in 1950 but all seventeen major fishing areas have either reached or exceeded their natural limits, and most are in serious decline.[16] The cod fishery off the coast of Newfoundland was the most recent of the big commercial fisheries to collapse.[17] In recent years the world annual fish catch has fluctuated around 90 million tonnes, and greatly increased fishing effort and expense has not led to bigger catches. More than 20 per cent of fish supplies now come from aquaculture, for which mangrove forests are often destroyed to make fish ponds—in which the fish also have to be fed.[18]

For 200 years the spectre of Thomas Malthus has challenged us to show that we can keep feeding ourselves. Given that the same challenge confronts even the lowliest of species, it isn't much of an aspiration for humanity, and insofar as it has diverted attention from the *consequences* of feeding more and more people, one might wish that the Reverend Malthus had kept his thoughts to himself. Human numbers are at present growing at 1.3 per cent a year— which is adding 78 million people to the world's total every twelve months, mainly in the poorer countries.[19] This amounts to the combined populations of the British Isles and Australia and is the equivalent of adding, every four weeks, the entire world population of the Neolithic. Clearly, with this level of change something else must change in response, but then something is always changing. In the shorter term, it could be a combination of human diet, human mortality and food production. Or only one of these.

SIMPLIFYING THE EARTH

Energy continuously flows through the living world in one direction, entering the system as rays from the sun to be channelled
through the life processes of tens of millions of different species
until it is dissipated as heat—the disordered movement of molecules. The sun is slowly running down, as it must, but life presents
the illusion of running against the second law of thermodynamics.
Life seems to wind itself up. With sunbeams and a few chemicals,
life builds and maintains temporary order of enormous
complexity in every individual. Countless lives of thousands of
species can be involved in the baffling dynamics of a whole
ecosystem such as a coral reef. Darwin called it 'wedging', this
tight packing of species, and he marvelled at it.

If two different animals or plants had exactly the same requirements in life it is unlikely that both could live together; eventually,
one would outbreed and exclude the other. Either that, or they
would come to an accommodation by evolving just enough ecological separation to enable both groups to survive. Perhaps a slightly
different food preference and breeding season would do the trick
if both populations survived long enough for these changes to be
favoured through natural selection. So it has come about that in
ancient ecosystems the niches are tightly packed but with every
species constantly pushing its envelope, so to speak, in an evolutionary fine-tuning of relationships. If one type is removed, the gap
will be filled by neighbouring species whose niches will have been
fractionally widened. If a completely new species is introduced there
may be no niche for it to occupy and the introduction will fail.
Occasionally the newcomer, having evolved in a different set of
relationships, may be able to outperform the established types, and
we then say that a new pest has been introduced.

Where the intention is to grow crops, the aim is usually to
replace natural complexity with a simplified system, or to eliminate diversity entirely. A thriving cornfield, in terms of its
diversity, is ecologically destitute above ground because almost
everything except corn is a pest or a weed to be banished. But
when one or more pests gets out of control, then the cornfield
provides a bonanza which could not have been better designed
for the pest's benefit because there is no competition or opposi-
tion from other species. Crop monocultures are notoriously

vulnerable to pests, so traditional systems were never based on vast areas of monoculture; but neither were traditional systems capable of feeding six thousand million people.

At present, despite more land having been brought under the plough, total arable land per head of population has fallen from over half a hectare in 1950 to well under 0.3 hectares now.[20] Authorities seem unable to agree about the significance of the millions of hectares of land that disappear every year under the expansions of roads, industry and residential areas. The disagreements exist partly because the agricultural value of the land is so varied and partly because nobody can be absolutely certain about future needs. There are other reasons for disagreement, which we shall turn to in the last section.

LIFESTYLES AND ENERGY

In Europe, North America, Japan, Australia, Latin America and the Caribbean, at least 75 people in every 100 live in towns. Globally, nearly half the world's population is urbanised, and the proportion is fast-growing. There are now hundreds of cities with more than a million inhabitants, and some megacities have more than 15 million inhabitants.[21]

As for the commercial energy we use, if hunter–gatherers needed about 2000 kilocalories a day to cook and keep warm, then the average Indian now uses six or seven times as much; the average Briton or Japanese uses about 50 times as much, and the average American gets through over 100 times more non-food energy than our ancestors did.[22]

Most commercial energy still comes from the photosynthesis of the past, as coal, oil and natural gas. Nuclear energy makes up most of the balance. Developing countries devote most commercial energy to industry, whereas a much larger proportion in the USA and Europe is used for transportation.[23] Of course there is enormous variation between nations, and the poorest billion people scarcely use any commercial energy at all, but as a generalisation it is fair to say that in every part of the world people are becoming western-style consumers to the extent that they can afford it.

Needless to say, there is a prodigious amount of waste from all this conversion of energy, and it takes the form of solids, liquids and gases. The bigger the town, the bigger the disposal problems. No such problems existed until the most recent 1 per cent of human history, for at low densities of population all human waste was quickly broken down in soil and water. One family could dump all its untreated waste into a river and another family ten miles downstream could safely draw water for drinking. Replace the two families with two towns and the river couldn't cope with even the inorganic waste, let alone raw sewage.[24] In crowded urban centres, sewage disposal alone presents a major challenge to civic authorities and many millions of people live without the most basic facilities, commonly at densities of around 10 000 people to a square kilometre. In the slums of Mexico City the pork tapeworm, which normally uses the pig and the human as alternate hosts, is now more commonly transmitted directly from human faeces to human mouths by contaminated river water or on vegetables.[25]

High-density living provided new evolutionary opportunities for old disease organisms, and that was bad enough. More recent problems and threats have resulted from industrial and agricultural pollution, and these have become truly global. Not only are they pervasive and difficult to control, they are relatively new to science so the health implications are poorly understood.

POLLUTION, DEGRADATION AND CONFUSION

With 80 per cent of commercial energy being used by the wealthiest fifth of the population it isn't unreasonable that big business attracts most criticism from the Green movement. But we must also accept that poverty itself is another 'major global scourge'.[26] The fact is that some forms of pollution and destruction are distinctly associated with affluence, while others are caused by poverty. Still other pressures can be associated with every stage between. We have space in the remainder of this chapter only to elaborate a little on these points.

Pollution

By pollution I mean the adverse effects of adding something new to the environment, or increasing the levels of something already there. Local pollution is most closely linked to poverty and under-development and is characteristic of big-city squalor, where people are obliged to live in accumulated waste because there is no adequate way of disposing of it.

Regional pollution often results from a higher level of affluence and is typically caused by traffic exhausts and emissions from industry. Choking smogs and toxic waterways are all too common in rapidly industrialising parts of the world, where more immediate development goals take priority over environmental standards. But regional pollution is not contained by national boundaries, and it can be caused by multinational companies as well as regional ones. In 1984 the Bhopal tragedy in India killed some 3500 people. It was caused by a gas leak from a pesticide plant that was an Indian subsidiary of an American-owned company. At the same time the huge Indian pesticide industry was growing at five to six times the global average, as it increased exports and supported the Green Revolution by annually producing tens of thousands of tonnes of pesticides, including DDT—which had long been banned for general use in the USA and most European countries.[27]

As a persistent chemical, DDT can blur the distinction between regional and global pollution. From soil it can drain into rivers and so reach the sea. Because it is stored in the fat of living organisms it can travel across entire oceans in sea creatures, and it steadily increases in concentration in those animals which are higher in the food chains because they collect it from the tissues of their prey. The process is called biomagnification, and it can produce concentrations tens of thousands of times higher than the background levels in water.

True global pollution by greenhouse gases is mainly the result of burning fossil fuels all over the world but, as the wealthiest people are doing most of the burning, this is still very clearly a problem of affluence.[28] The same can be said about the release of chlorofluorocarbons (CFCs), which are destroying ozone in the stratosphere. If the richest nations reduce their emissions substantially, as they must, and the most populous industrialising countries increase theirs only moderately, the picture will change. More smoke and gas can obviously result from a great many people each producing a little than from far fewer people each producing rather more. This simple truth can be found in the well-worked equation: Impact = Population × Affluence × Technology.[29] The same formula, which has its limitations, indicates the need for helping developing nations to bypass the old dirty and wasteful technologies and get directly into the cleaner, more efficient versions—a strategy that has become known in development circles as leapfrogging.

Environmental degradation

Degradation is sometimes a consequence of pollution, and the effects of pollution can be worse in a degraded environment, but again I believe it makes for clearer thinking if we maintain distinctions. By degradation I refer mainly to environmental losses rather than the additions that cause pollution.

Losses of soil, water, forest cover and wild species are commonly caused by rural peasants who have to scratch out a living wherever they can, and whose capacity for pollution is minimal, but similar losses are caused by mechanised farmers, urban sprawl, company executives and all levels of government in both capitalist and planned economies. In Ethiopia I have seen highland subsistence farmers trying to plough patches of rubble from which virtually all soil had long since washed away. In Queensland, Australia, contractors with bulldozers and massive chains are at present clearing eucalyptus woodland in order to bring more marginal land into production. As communities grow and standards of living and expectations rise, yesterday's low priorities become today's economic necessities, or so it will be claimed. The impacts of poverty and big business are not always separate. In the forests of Asia and the South Pacific loggers open up the forest, and when the logging is over subsistence farmers can be expected to move in, clearing, burning, cultivating and ruling out any chance of forest recovery.

Loss of tree cover has far-reaching consequences. Forest cover on mountains acts as a sponge which intercepts the rain and slows its return to the sea. Water from denuded mountains crashes violently into rivers, leaving the headwaters to run dry and contributing to bigger, faster floods further down. More subtle effects of tree clearance can affect prime agricultural land. More than a million hectares in Australia are being turned into salt flats largely because of the removal of deep-rooted trees. This has allowed salty water to come to the surface, where salt remains when water evaporates. The same thing has happened, but more quickly, on irrigated croplands where salt was at first so deep that it wasn't an issue.[30] The problem is by no means unique to Australia: in the USA more than five million hectares were salt-damaged by the mid-1980s.[31] Nor is land degradation anything new; salinity contributed to the decline of the first city-states in Mesopotamia. In those days, communities could move on to new regions but now there is no unoccupied farming land of any quality to move to.

It may take a practised eye to notice the early stages of soil erosion or salinisation, but water depletion soon shows up. About 70 per cent of all the water we use goes into irrigation because, depending on the crop and the climate, it takes 600–1600 litres of water to produce one kilogram of grain.[32] There is no overall shortage of fresh water, but supplies and needs around the world don't match up at all well. As a result, rivers are being overused and water tables are falling, quite dramatically in some places. China's great Yellow River failed to reach the sea for the first time in 1972; and in 1996 it didn't even make it as far as Shandong Province, the last province through which it normally flows.[33] America's Colorado River now peters out in the Arizona Desert, and even Australia's Snowy River is down to one hundredth of its former flow because of the demand for hydroelectric power and irrigation water. As I write this, the irrigators are demanding the release of more water from the hydro scheme; the need for urban energy is conflicting with the demands of food production.

As for water in the ground, the levels are falling because withdrawals are exceeding the rate of replenishment. This applies to an estimated one-fifth of irrigated cropland in the USA and is widespread in India, where withdrawals are estimated to be twice the rate of recharge and water tables are falling by 1–3 metres a year.[34] A similar rate of fall has been reported in northern China, home to about 100 million people, where the water table has fallen by 30 metres in as many years.[35]

Humans and other species

Nobody knows how many species there are in the world, but 12.5 million has been suggested as a conservative working figure.[36] The most species-rich systems appear to be the long-established ones, such as tropical moist forests and coral reefs, where tight niche-packing has had longest to evolve. The extinction rate can be high when species-rich systems are converted to human use or otherwise destroyed, and habitat modification is by far the major cause of present-day extinctions. There have been mass extinctions in the past, but 'sudden' extinctions in geological time were not at all sudden in the human time scale.

The tragedy of a sudden extinction is that it wipes out millions of years of accumulated hereditary information. The digital record in DNA comes to an abrupt end and no further evolution is possible from that lineage. The late Julian Simon, who until 1998 was professor of business administration at the University of Maryland, had a lot to say in favour of ever-bigger human populations but was obviously in unfamiliar territory on the subject of extinctions. 'The extinction of some species' he wrote with his co-author, a political scientist, 'is an essential precondition of the development of newer and better versions'.[37] No it is not. Where are the new and better versions supposed to come from? What if the first *Homo* line had been nipped in the bud?

Humans are leaving precious little room for competing animals over much of the Earth's surface. A small minority of species, mainly the ones we call pests, have done very well out of us in that they have been able to exploit our crops and buildings. A few favoured species, such as garden birds, may also have benefited in some countries, and birds as a group are favoured by conservation movements. But even among birds many specialised types are threatened with extinction and many more have been displaced by intensive farming. In Australia's wheatbelt, 60 per cent of all species are in decline.[38] Large mammals, which are a threat to livestock or human life, have been eradicated from regions that have long been settled, and they are under intense pressure elsewhere. Wildlife can sometimes be best conserved when there is money to be made by protecting habitat instead of converting it to other uses. Tourism, controlled harvesting of wild species and sport hunting can all be a basis for profitable protection in some circumstances.[39]

Human populations, at *all* levels of wealth and technology, have a devastating impact on other species and their habitats. Whereas pollution tends to be cut back to some extent by the most affluent nations, the pressure on most animal groups continues to intensify with economic activity.[40] The mainstay of habitat protection is now a global network of national parks and other reserves covering some 6 per cent of the land's surface, though only about half of that area can be considered to be totally protected, and that includes the enormous contribution of the Greenland National Park.[41] At least three times as much land would be needed 'to constitute a representative sample of the Earth's ecosystems'.[42] Some quite famous protected areas around the world, especially in Africa and Asia, are giving way to inexorable pressure from rural communities whose genuine agricultural needs are in conflict with wildlife conservation.[43]

There is now international collaboration in trying to preserve genetic diversity in seed banks and zoos, but seeds don't remain viable indefinitely and progressive zoos cost a fortune to run and are dependent on gate money from the public. 'Frozen zoos', in the form of deep-frozen genetic material from species threatened with extinction, may be a last-ditch conservation strategy for selected animals.[44]

WHO CARES, AND WHY?

Where people have been given the opportunity for family planning the reduction in average family size has sometimes been remarkable, though the reasons for the so-called demographic transition to smaller families are not as obvious as was once thought. Whereas Brazil and South Korea moved swiftly to near-replacement-level fertility rates, North America's fertility rate is slightly higher now than it was in the 1970s. Contrary to earlier wisdom there is 'no tight statistical link between development indicators and fertility rates', but the transition, once it has begun, proceeds faster with higher levels of development.[45] Nor are long-term trends easily predictable: in England and Wales in the 1930s average family size was down to two, and it was expected that the United Kingdom would have no more than 44 million people by 1970, by which time, it was feared, one in six people would be over the age of 65. In fact, in 1970 there were 55 million people, and twice as many were under fifteen as were over 65.[46]

Predictions and projections have no place in this chapter, but most opinions and attitudes in the polarised debate about population and environment do seem to be shaped at least as much by visions of the future as by the existing state of affairs. Leaving aside the dogmas of religion, and accepting that political Greens come in all shades, those with the most to say in the debate tend to be concerned environmentalists on the one side, and unconcerned economists together with some social scientists on the opposing side. The above-mentioned Julian Simon was for years a champion in the ranks of the latter. Paul Ehrlich, Professor of Biological Sciences at Stanford University, has been a well-known doomsayer on the environmentalist side since he wrote *The Population Bomb* in 1968.

In focusing public attention on the state of the world's environment, the UN Earth Summit in Rio de Janeiro was a milestone. It was that conference in 1992 which produced the 40-chapter action plan called Agenda 21. The institutions and intentions that flowed from Agenda 21 are still functioning and there have been some substantial achievements, especially by non-government organisations, but I haven't heard of anybody getting overexcited at the rate of progress.

Destruction of the remaining wild places, with the resultant loss of diversity, is a particularly intractable problem because of the impact of population density at all levels of wealth and technology. I call it a problem, but those who are concerned about other species have a hard time trying to impress people who couldn't care less. This is because no matter how objective scientists try to be in assessing ecological wellbeing, questions of *concern* are a matter of feelings and perceptions as well as scientific data. Put simply, public concerns are the issues that people get concerned about. One person's passion is another person's yawn.

Still, the case for diversity does include purely utilitarian arguments as well as appeals to ethics and aesthetics. Billion-dollar industries are based on medicines that were originally discovered in plants—aspirin from willow bark and penicillin from fungi, for example. Unexpected discoveries are being made all the time, and not only in plants: invertebrates are now being tested for antibiotics, and frogs have already contributed to the treatment of peptic ulcers.[47] In agriculture, more than a thousand new rice varieties have been launched since 1966 in order to maintain disease resistance in the main commercial types.[48] All the genes had to come from somewhere and, as most species have never been formally tested, nobody knows what we are losing as millions of genes disappear. Geneticists can only manipulate genes, they can't make them, although the people who are least convinced by scientists' concerns often seem to be the most confident that science will keep coming up with technical fixes.

Among academics there is still a great divide between those with natural science backgrounds and those with entirely different interests. Ecologists have always had population interactions at the core of their subject, but economists see human populations as *markets* in which people are *consumers* so the topic is approached from an entirely different direction. Some social scientists arrive on the scene from directions that are more obscure. One sociologist, Frank Furedi of the University of Kent in England, has adopted the theme that environmental scientists who are concerned about population pressure are really just part of a 'population control lobby' which is searching for 'excuses' to justify its unstated agenda. This agenda, according to Furedi, has to do with power politics, and isn't based on concern for the natural environment at all.

Accepting that this is a genuine opinion, it is not a view based on much ecological understanding. At one point in his book, arguing that environments can collapse if they are not cared for by people, Furedi cites the observation that terraced hillsides and plantations in Nepal deteriorated when the farmers were absent.[49] Well, they would do—they were made by people in the first place. The distinction that Furedi fails to make here is that between natural forest cover and a landscaped garden. Or a cabbage patch for that matter.

One doesn't need to be an E.O. Wilson to see the need for more overlap between the academic disciplines. As a priority, there does seem to be a solid case for ecologists to learn the language of economics, and for social scientists to be given a good dose of biology. Urban communities now live at such a remove from their ecology that children of school age in western cities have to be taught that dairy products come from cows. Some form of economics must obviously mediate between ecology and the functioning of modern societies, but the mediation so far has been very one-sided. In the end, however, ecology must have its due. I can say this because, for all the sophistication and wizardry in our new world of metal, plastic, glass and concrete, we are no less dependent than we ever were on that ancient, silent chemistry that catches the energy of sunshine and holds it captive in leaves and shoots and stems.

CHAPTER 12

RECIPES AND OUTCOMES

W HEN FACTS TAKE MANY decades to uncover, and publicity centres on the interesting snippets, there is bound to be confusion and some reversals in public perception and understanding. One writer claims that genes are destiny; another argues that this is rubbish. One reporter announces the discovery of a 'God gene', another says no such claim was ever made. Controversy sells. In this chapter I shall stay mainly with the mainstream scientists who are trying to unravel the relationships between genes and the visible person.

We saw in Chapter 2 that humans have tens of thousands of genes scattered among some three billion base pairs of DNA, which is bundled in 24 pairs of chromosomes. Typically, about 3000 base pairs make up a gene, but there is a lot of variation. Finding all these genes involves working through the DNA of every chromosome and writing out the sequence of the three billion bases—A-G-T-C—as they occur. The triplets of bases (codons) which specify amino acids must be recorded together with the bases that act as punctuation marks at the end of polypeptide runs. The only way to find every human gene is to work systematically through the entire genome of three thousand million base pairs.

READING THE COOKERY BOOK

In the mid-1980s scientists began to think seriously about this monumental task, and the US Department of Energy (DOE) was among the prime movers. With responsibility for nuclear energy, the DOE had a special interest in the DNA molecule, particularly in the way it responds to ionising radiation. Various health departments and commercial laboratories around the world could also see the value of one day having the complete sequence of the human genome. It took a few years to get a plan off the ground, but in 1990 the Human Genome Project (HGP) was officially launched as an international research effort. Major sequencing laboratories are located in the USA and the UK, with other centres in Germany, France and Japan, but there are participating and interested scientists in many countries, linked through the Human Genome Organization (HUGO), which was set up in Switzerland in 1989.[1]

The task of sequencing the entire length of DNA is not done by starting at one end of a chromosome and working through to the other. Rather it proceeds by mapping the chromosomes at successively smaller scales, starting with large fragments and then mapping the fragments in greater detail to produce working drafts. Eventually, gaps are filled and overlaps are sorted out until a finished, accurate sequence is reached. The tools for this sort of work include enzymes (restriction enzymes), which are so specific that they cut the DNA strand only at sites where a few bases are in a particular sequence—such as GAATTC. DNA fragments can be sorted for size by floating them in a gel and causing them to move through it with an electric charge. Smaller bits move faster. Fragments are cloned so that scientists have thousands of copies to work with. So small is this material that with an ordinary light microscope you cannot even see the fragments.

When the project began, a single researcher with the best available technology needed about a year to produce a continuous sequence of 20 000 to 50 000 base pairs, at an average cost of US$5 per base pair. The project looked set to last for about fifteen years. But computerised sequencing machines and other technological advances brought costs down to 20 or 30 cents per base pair and the target date for completion has been brought forward to the year 2003—the fiftieth anniversary of Watson and Crick's famous description of the DNA molecule. A 'working draft' of the genome was produced in June 2000, by which time 38 000 genes had been identified.

The original DNA material was obtained from human sperm and blood. Sperm contains the full set of chromosomes but blood collected from females lacks the male (Y) chromosome. The material used is all from anonymous donors, so the fully sequenced genome will represent a mix from a number of unknown people—a sort of generalised person. To find all the different alleles of particular genes it will be necessary to sequence many parts of the genome many more times and from different donors. Even then, science will have taken only the essential first step in working out the recipe for our species. Geneticists will have a list of all the known genes, but that still won't explain how the genes work in the living animal—that is, how they are all able to interact in a controlled and regulated way. This is where the real work begins and, according to one writer, it could generate at least 50 000 professorships.[2]

GENES AT WORK

Geneticists are already familiar with at least some of the ways in which genes are orchestrated: the mechanisms by which they are timed to switch on, do their thing, and switch off again in an interactive manner. In Chapter 2 we saw that when DNA is being copied during cell division, another nucleic acid, messenger RNA, forms a template in a process called transcription. Control of gene activity is typically exercised at the transcription stage by 'regulatory genes', which can promote or repress the action of other genes. Control of gene expression can also come into play after transcription, when regulatory proteins influence how the RNA transcript is translated.[3]

The task of finding out how the genome actually functions would scarcely be approachable if our whole genetic make-up was exclusively human, because to understand how we work at the molecular level it is necessary to interfere with normal processing. Experimental changes have to be made in order to see what happens, and for that we need experimental subjects.

There is precious little difference between the genome of a chimp and that of a human because we have been evolving separately for only a few million years. The differences are greater the longer organisms have been following separate evolutionary paths, but all genomes should have something in common if they all evolved from the same cells to begin with. All surviving types have obviously been successful, so they must have an unbroken line back to a common form of life; natural selection has no way of scrapping successful genes and starting afresh. Genes for basic life processes have been conserved since they first became hereditary units, which is why we find ourselves sharing genes with mice and worms and tomatoes as well as with other primates. You can make all kinds of buildings with the same old bricks. It is because of this continuity of life that the genomes of comparatively simple species, such as the nematode worm that we met in Chapter 2, are important to the Human Genome Project.

It has been estimated that defective genes account directly for some 4000 human ailments, many of them very rare; but for the most part genes have complicated relationships, and the functioning of our genome is still an incalculable puzzle. Gradually we shall come to know which of our genes does what, in combination with which others, and how the functioning is controlled at the molecular level. And we shall acquire this knowledge largely from experiments with other species and by switching genes between species. There are already well-established strains of laboratory mice whose genomes are partly human. Even then, it is one thing to know how our genes function in a mouse and another thing to know how they operate in a human being. And it is one thing to know how the human genome functions in a generalised person and another to understand how small variations in that genome can account for six billion people all being different. Genomes of individuals will be a future reality.

Nature and Nurture

If we took a few dozen cuttings from the same bush they would all be genetically identical. If we planted them all in one bed, under conditions as uniform as we could make them, we could expect the rooted cuttings to grow equally well or equally badly depending on how we cared for them. As there would be no genetic variation, any slight individual differences would be attributable to differences in the treatment; perhaps the bed wasn't quite as uniform as we thought.

Now suppose we took our cuttings from a few dozen bushes of the same species which we found growing in very different localities. Let's say they were genetically varied and nicely adapted to their different habitats. This time, if we planted all the cuttings in uniform conditions we could not expect them to do equally well. Some might do much better than others and it would be reasonable to attribute most of the variation to genetics. Finally, suppose that, instead of planting our mixed bag of cuttings in a uniform bed, we planted them randomly all over the place. Once again, some would do a lot better than others, but now we wouldn't know how to account for the differences because a variety of genes would be responding to a variety of environments. It wouldn't be possible under these circumstances to separate the influences of one from the other. Generally speaking, this is how it is with people.

Every individual carries a selection of the species' genes. That is the individual's genotype. When the individual is fully mature those genes will have expressed themselves in response to all the signals they received, to produce the developed organism that we see and call the phenotype. It is easy to see how identical genotypes can produce different phenotypes: cuttings from the same bush will not grow the same way in different parts of the garden and identical twins will not be look-alikes if one is starving and the other well-nourished. On the other hand, in complex animals, different genotypes will never produce identical phenotypes no matter how equally they are treated. On the contrary, if equality is what you want, you might be able to level them up to some extent if you know how best to treat them *unequally*.

From a gene's eye view, 'the environment' includes everything not coded in DNA that affects gene expression. The environment of a plant includes all the outside influences on its growth, and as plants grow they can change the external environment profoundly; a forest clearing is transformed by the bushes and trees that successively cover it. Animals also change their environments for themselves and each other, so the gene–environment interaction, or nature–nurture relationship, is a two-way process. But the world of mammals is more complicated because the environment includes the inside of the womb during the gestation period, the birth and rearing circumstances, and, in social mammals, the wider social environment. Furthermore, the phenotype of an animal includes its temperament and other mental attributes as well as its physical characteristics, so the interactions are far-reaching. If human parents are predisposed to be shy and withdrawn, for instance, they are likely to set that example in their children's environment as well as passing on the predisposing genes.

On an individual basis it may never be possible to separate the relative contributions of nature and nurture to a person's development, and some would say it makes no sense to think about it. Certainly the old arguments about nature *versus* nurture were misconceived—rather like seeing the ingredients of a cake pitted *against* the cooking. But it does make sense to ask to what extent different ingredients can produce different cakes when they are all cooked the same way. That line of questioning is sound because genes can account for particular variations in populations. But with genes and people the answers are hard to get at because we can't manipulate human populations in an experimental way. The genotypes are all different and they are never expressed in exactly the same environment. Different ingredients are getting different cooking.

HERITABILITY AND THE VALUE OF TWINS

If you measured the width of everyone's head in your district you would get a range of dimensions from the narrowest to the broadest, with most sizes probably being somewhere in the middle. If this variation had nothing to do with genetics but could all be accounted for by environmental factors, such as whether people had worn bonnets as babies, then heritability for head width would be zero. If, on the other hand, environmental factors had made no difference and the variation was all the result of genes, then heritability for head width would be 100 per cent. Heritability has a statistical definition, but roughly, for any trait in a given population, it is a measure of how far the phenotypic variation is caused by genetic variation. It follows that for any trait, the more equal the environment, the more the variation in phenotypes must be caused by genes. If there is no genetic variation, as in the number of heads or ears per person, then it makes no sense to talk of heritability even though the body plan is written in the genes.

In domestic animals, heritabilities are found during selective breeding for particular traits such as milk yield, growth rate or egg production. The figures are worked out by comparing averages for the traits in offspring with those of parents in the selected line and with the wider population of animals not being selectively bred. The higher the heritability, the faster the gain through selective breeding. The heritability for weight of cattle at slaughter age has been put at 85 per cent, that for egg weight in poultry at 60 per cent. Milk production in cattle came out at only 30 per cent and that for egg production in poultry at only 20 per cent, but this is more than enough for us to have quickly developed egg-laying breeds that are several times more productive than birds bred for meat.[4]

In case anyone is wondering, heritability for head breadth in humans is said to be 95 per cent.[5] But the figure wasn't found by selective breeding.

Identical twins are produced when a fertilised egg separates into two bundles of cells (zygotes) soon after cell divisions begin. If the two zygotes remain joined at any point, then so-called Siamese twins will be born—named after the famous pair, Chang and Eng Bunker, born in Siam (now Thailand) in 1911. Though joined at the chest and never separated, the boys survived and as adults they married and fathered a total of 21 children.[6] Technically, identical twins are said to be monozygotic (MZ for short). They have exactly the same genotype, unless mutations occur during development, so they are always the same sex. Non-identical or fraternal twins (dizygotic, or DZ) develop from different eggs and sperms so they have about half their genes in common, as do all brothers and sisters.[7] Most of what we know about the heritability of human traits comes from studies of twins. As Francis Galton realised in the nineteenth century, they provide natural experimental designs that could not otherwise be arranged for humans. It was Galton, too, who first worked out the mathematics for revealing correlations.

There are various ways of comparing twins in heritability studies. If a trait is highly heritable, then identical twins raised together should be more alike than fraternal twins raised together, so one method is to find examples of twins of both kinds and measure the pairs, trait by trait, to see how well they correlate. Researchers also have to take into account the fact that twins have shared the uterus during gestation, so they have had a more similar environment before birth than did other siblings (most identical twins actually share a placenta). Another problem is the possibility that identical twins, because they were more alike to start with, will have received more similar treatment as they grew up—making their environment more similar than that for fraternal twins.

An approach that avoids the criticism of similar treatment is to find identical twins who have been separately adopted and fostered in different homes and see if they still remain very similar. If they do, and if they are even more alike than are pairs of fraternal twins reared together, then the case for a genetic explanation is very strong indeed. The initial problem here is that of finding enough cases to make up a good sample.

TWINS, FRAUDS AND STRONG FEELINGS

Following Galton, the outstanding figure in early nature—nurture studies based on twins was the British psychologist Cyril Burt (1883–1971). Professor Burt was involved in developing statistical techniques used in intelligence testing, and he firmly believed that intelligence was largely inherited. In 1955 he claimed to have studied 21 pairs of separated identical twins and found a high correlation for intelligence. He added more twins to his study in subsequent years and they all served to confirm his earlier findings, but in 1972 his work was called into question by Leon Kamin, a psychologist at Harvard who disagreed with everything that Burt had stood for. Three years later, after studying Burt's data very closely, Kamin announced that: 'The numbers left behind by Professor Burt are simply not worthy of our current scientific attention'.[8] Former supporters of Burt also found fault with the numbers, but Burt's reputation was not finally destroyed until his biographer, in 1979, cast doubts on the professor's mental health and conceded that in later life Burt had been falsifying his records.[9]

Burt's fellow scientists had apparently held him in such esteem that for years they had failed to give his work the critical scrutiny that scientists can't usually avoid. A more recent review of the Burt affair has raised doubts about the biographer's findings, but Burt's reputation can never be restored and his twin studies are no longer included in published research.[10]

Full marks to Leon Kamin for his scientific vigilance; I doubt that he would want any credit for helping to give new respectability to the study of twins. Leon Kamin, together with British biochemist Steven Rose, Harvard geneticist Richard Lewontin, and Stephen Gould, whom we met in Chapter 6, have for decades been among those united in playing down the importance of everything biological that they think might be used against their brand of politics. Rose, Lewontin and Kamin, collaboratively writing *Not in Our Genes* during the early 1980s, explain their association with the 'radical science movement' and devote 300 pages to countering the alleged biological reductionism and determinism of the political New Right as exemplified, they say, by the likes of Richard Dawkins and Edward Wilson. Rose and his colleagues don't see themselves as *environmental* reductionists or determinists because they don't deny the importance of biology in human nature. Yet they don't accept that society is in any way predictable on the basis of interactions between genes and culture either. They claim that feedback between organism and environment goes beyond mere interactions when the former actually reorganises the latter.[11] As for the heritability of intelligence, they state flatly that, with or without Cyril Burt, twin studies as a whole cannot be taken as evidence for the heritability of intelligence.[12] Despite the jargon and the sweeping statements ('"Science" is the ultimate legitimator of bourgeois ideology'), *Not in Our Genes* does present familiar arguments against biological determinism, and I believe most biologists (including Wilson and Dawkins) could find plenty of common ground because nobody seriously believes that we are all gene-driven robots. But Rose and his comrades scarcely mention the pitfalls of environmental determinism or the many writers on their side of politics who fall into them. Nor do they offer any new approach to the practical study of nature–nurture problems. They criticise the findings from studies of twins but don't suggest improved methods. They have other priorities, and they quote Karl Marx in saying: 'The philosophers have only interpreted the world in various ways; the point, however, is to change it'. Be that as it may, most research biologists are not campaigning for radical change in society, though they will surely be pleased if their work can be helpful to medical science. All researchers, however, must be hoping to gain a bit more specialist understanding. That much, at least, is what thousands of scientists are expecting from their daily routines in the laboratories of the Human Genome Project, and from the continuing study of twins.

STUDIES OF TWINS AFTER BURT

Since 1979 there have been some important studies of twins in Europe and the USA, but the major program is the Minnesota Study of Twins Reared Apart (MISTRA), which is based at the University of Minnesota. Over the past 20 years it has documented about 1000 pairs of twins, including well over 100 sets of reared-apart twins and triplets from countries all over the world. And the documentation isn't just a matter of ticking off questionnaires: the participants spend a week at the study centre, expenses paid, subjecting themselves to about 50 hours of intensive medical and psychological testing.[13]

Measurements such as height, weight, blood pressure and brain waves are relatively straightforward, but with psychological assessments each trait is measured with more than one type of test. Multiple tests are used to measure intelligence, personality, temperament, leisure interests and social attitudes—right down to likes and dislikes on TV. The intelligence tests are of three kinds and are administered while twins are in separate rooms. Home environment is also categorised systematically, taking account not only of socioeconomic status but including cultural, educational and other possible influences. Checklists are completed for household items and facilities that were available as the twins grew up—things such as sporting goods, dictionaries, power tools and artwork.[14] Without computers the statistical analysis of all these data would be a nightmare.

Sometimes, even though they were separated at birth, identical twins develop the same habits and foibles to a degree that seems quite uncanny. One pair, having been separated as babies, didn't even know of each other's existence until they came face to face as grown men at a fire-fighters' convention. Not only did they look alike, down to the moustache and wire-framed glasses, they discovered that as well as having become fire-fighters they were both bachelors and both keen on hunting and motorbikes.[15] Nobody is suggesting that there are genes for fire-fighting, but when two people are psychologically very similar it must increase their chances of making the same decisions in life and so makes coincidence a lot more likely. Whole books have been written about the lives of MISTRA twins.[16]

So how similar are identical twins? Similarities are recorded as correlations. If every pair matches perfectly on a given measurement (extremely unlikely with a large sample of any living things) there is a perfect relationship and the correlation is expressed as 1. When they show no relationship the correlation is zero. Some correlations obtained from many pairs of identical twins, reared apart and together (latter in parentheses), are as follows: fingerprint patterns 0.96 (0.97), height 0.86 (0.93), weight 0.73 (0.83), intelligence 0.69 (0.88), personality variables 0.48 (0.49), occupational interests 0.40 (0.49).

Correlations for fraternal twins, whether raised apart or together, are invariably lower than for identical twins, and the differences are used to estimate heritabilities. The usual estimates range from around 40 or 50 per cent in personality and vocational interests to about 70 per cent for intelligence and 90 per cent for height. After reviewing all the data available, MISTRA researchers announced in 1990 that 'For almost every behavioural trait so far investigated, from reaction time to religiosity, an important fraction of the variation among people turns out to be associated with genetic variation'.[17] The researchers also recognise the two-way interaction between genes and environment and point out that children with different temperaments are likely to elicit different responses from their parents. Critics of twin studies have argued that identical twins are likely to get more similar treatment from parents than are other children. The MISTRA and other researchers don't disagree; they just say that because of their identical genomes the twins themselves help it to happen.

Older children and adults can be expected to seek out for themselves the social and other environments that they prefer, and because adults are freer to make decisions than are children, we might expect coincidences to accumulate as twins grow up. Older identical twins, reared apart, get even higher scores for similarity than do younger ones. The same reasoning could explain why ordinary siblings who have been reared in the same home are *not* more alike as adults than are siblings who were reared in adoptive homes. It is as if the home influence fades as the different genotypes are progressively expressed in the wider world.

None of this means that parents have no influence on such traits as social attitudes in their children, and of course gross deprivation or mistreatment would obviously have an impact. But the evidence from twins does indicate that, on average, we are no more like the brothers and sisters with whom we grew up than we would have been if we had been reared in different foster homes.[18] While some people may find this surprising or disappointing, others might take heart from the implication that conscientious parents need not feel that they have somehow treated their offspring unfairly just because the children turn out to have quite different characters. Different characters and abilities are to be expected from different genotypes.

INTELLIGENCE

People have wondered about human mental faculties since the ancient Greeks, but modern ideas of intelligence as something that could be measured have developed from the early work of Francis Galton, which was furthered at London University by Charles Spearman (1863–1945) and Cyril Burt. It was Spearman who introduced the idea of a general capacity for learning and problem-solving which correlated highly with separate mental abilities. It came to be known as Spearman's g. Some researchers later claimed that there was no single g factor but rather a collection of dozens of abilities that were all independent. There are now whole libraries on the subject of intelligence but the dominant position is based on something like the following.

There are really three different concepts of intelligence. First, there is the biological basis for what makes one person different from another in mental agility. We can refer to this as intelligence A. There may be any number of reasons why one healthy, normal brain should perform differently from another, but the biological basis of brain power is still largely a mystery despite the enormous individual differences that exist.

Next, there is the real-life mental functioning that comes with personal development, and which shows in our everyday performance whenever we need to use our wits. This we can call intelligence B and it obviously counts in many of life's endeavours and achievements. The trouble is that, although we can easily recognise it, there is no way to measure it or compare it in people unless it can be represented by some sort of standardised test. Intelligence C, or measured intelligence, is that ability, whatever it is, which standardised tests measure when they are used to reveal an intelligence quotient, or IQ.[19]

One could protest (and many do) that in everyday life we use our brains in a great many ways that are not apparent in the tests for IQ. But this raises a point that is not usually discussed in this context: all vertebrates use their brains in ways that are not assessed in IQ tests. If they were, then all sorts of species would get higher scores than would humans. For example, if we chose to include tests of visual discrimination and certain kinds of memory then I, for one, would do very badly compared with many birds and possibly even some reptiles. A bird of the crow family called Clark's nutcracker is believed to locate more than half of the tens of thousands of pine seeds it has buried in the woods. It digs them up over the course of several months, often after snow has fallen, and it doesn't come across them accidentally: this bird recognises where to look for them even though it has caches in thousands of places.[20] I once had trouble retrieving two dozen small mammal traps from the woods after one night, and it hadn't been snowing! It can be hard enough keeping track of the car keys. And think of the seabird that flies back from miles away and unerringly returns to a camouflaged egg among the millions of similar-looking pebbles on the beach. Some of us have difficulty finding where we parked the car outside a supermarket. Those who would extend our concept of intelligence to include more non-reasoning components can only support a view that I have long held: many animals have more sense than we give them credit for.

Intelligence tests

Reasoning, language, and numerical skills are the special characteristics of humans everywhere. Modern intelligence tests, such as the most recent version of the Weschler Intelligence Scale for Children (WISC-III), include sets of items that call for different kinds of ability—verbal, numerical and reasoning. Reasoning can be demonstrated without words or numbers when the puzzles are pictures or shapes.[21] The subsets are scored separately but an overall score is obtained to give the IQ. People rarely do equally well on all the different subtests: some are better with words, others with the spatial items; but people who score highly on one do tend to be better than average on the others, which is why Spearman came up with his g factor. Today, specialists who accept the existence of *g* still argue about what it might be, and about its importance.

Obviously, an IQ has practical value only to the extent that it stands as a true and reliable indicator of intelligence B, so after nearly a century of test development one would hope to find high correlations between scores on modern IQ tests and what we think of as evidence of intelligence in everyday life—especially academic achievements. This is indeed the case, but it wasn't like that to begin with. In the early days there were false assumptions about intelligence, some badly designed tests and some quite improper applications of them.[22]

It was once thought that intelligence was a sort of fixed capacity that we were born with. We now know that it is impossible to separate genetic potential from personal experience, as the one cannot develop without the other. If all children have a favourable, and equally favourable, environment in which to develop intelligence then the full variation in human intelligence can emerge. Intelligence as we measure it will not develop properly if children don't have access to adequate schooling. The initial effects of intensive preschool inputs don't appear to last very well, but various adoption and other intervention studies have shown that intelligence will respond by several points of IQ to improved education—if schooling was seriously inadequate to begin with. Better schooling for everybody cannot be expected to close any gaps between individuals, however, because the brightest children in poor schools are likely to benefit just as much as the less scholarly children when circumstances are improved.[23]

Natural variation will always be with us; it is the stuff of evolution, and our brains could not have evolved without it. But why mental ability should vary as much as it does remains a mystery. It far exceeds the range of physical abilities. Our best athletes can run only two or three times faster than the average, or jump that much higher, but the mathematics genius can perform feats that are the equivalent of outrunning express trains and leaping over skyscrapers. Ability of that sort is totally unpredictable. The first Indian to be elected to the Royal Society was Ramanujan Srinivasa in 1918. He emerged from poverty in his homeland having acquired astounding mathematical insight without any training.

We know that measured intelligence continues to develop until adolescence and that it declines in old age. Because IQ stays remarkably constant between these periods, it is one of the most predictable psychological traits. Yet it appears that the average IQ has been increasing by about three points per decade since widespread testing began so that tests have to be re-standardised every so often. Those who score 100 on today's tests would probably have averaged about 106 on the tests of 20 years ago.[24] The phenomenon has not been explained but it is unlikely that actual brain power can have changed much. Possibly people in urban societies are getting more familiar with the sorts of items that appear in tests. No doubt we have become less aware in other ways. Try asking townsfolk what phase the moon is in and I doubt many could tell you, but I feel sure the average Neanderthal would have known.

The IQ itself is based on the assumption, never seriously challenged, that intelligence in any population has a normal distribution: that is to say that most individuals will be in the middle of the range, with progressively fewer people extending away from it, above and below the average. This is the usual pattern of variation throughout nature. You would see it by taking a few hundred people randomly off the streets and ranking them according to their height in increments of one centimetre. Get everybody in the same size category to stand in a queue with the front person toeing a white line. Arrange the queues so that the shortest people are at one end of the white line and the tallest are in a queue at the other end. Your longest queue would be fronted by the person of average height standing at the middle of the white line. Seen from above, this frequency distribution forms a bar graph, which, after smoothing out all the little steps, forms the bell-shaped curve of a normal distribution.

An IQ of 100 indicates average intelligence. It is the number assigned to the mean score for a large number of people on a test standardised for their age group. Because of the mathematics of a normal distribution we know that 68 per cent of people have IQs between 85 and 115, and 95 per cent of people have IQs between 70 and 130.[25] Statistically speaking, a person with a very high IQ is just as 'abnormal' as a person with a correspondingly low one. Like the very shortest and the very tallest individuals on the ends of our white line, there won't be many of them.

One of the most valuable applications of modern tests is in the diagnosis of learning difficulties. As the norms are based on the test results of large numbers of children, they indicate what can reasonably be expected. Children who fall behind in learning to read, for example, may have adequate IQs but are handicapped by specific learning difficulties which can be identified and remedied. Determining the IQ is an important step in the diagnosis but it must be done properly. For this reason, western education authorities set procedures and regulations for intelligence testing.

IQ, race and social issues

Except for abortion I can't think of a hotter potato in western societies than the issue of race and social status in relation to IQ. Passions were raised to new heights in 1994 by the publication of *The Bell Curve*.[26] This incendiary book by two American academics presents a mass of statistics based largely, though not entirely, on the National Longitudinal Survey of Youth. In that survey a carefully compiled sample of 12 686 young Americans had their IQs recorded in 1979 and then became part of a program that has been following their fortunes ever since. The authors of *The Bell Curve* drew attention to this and many other studies which have shown different ethnic groups to differ in average IQ.

Ashkenazi Jews are said to score highest, and they are certainly represented in conspicuously disproportionate numbers among the high achievers of professional and academic life. East Asians, principally people of Chinese and Japanese origin, are also overrepresented in American professional and academic circles, and several studies have shown them to have an average IQ a few points above that of white Americans. People of African origin, on the other hand, show an average IQ which is around 15 points below that for whites, and they are seriously underrepresented in the more academically demanding positions. In deliberately relating measured intelligence to social status, and then going on to discuss the wider implications of their claims, the authors of *The Bell Curve* went far beyond fact-finding and straight in at the deep end of social politics. They got all the publicity they could have wished for—or dreaded. Some notable critics then had their own responses published in a 1995 edited volume called *The Bell*

Curve Wars.[27] Reverberations of the exchange can still be found on the Internet, but a more useful spin-off was a summary of the present state of accepted knowledge about intelligence, published in 1996 by the American Psychological Association.[28] More recently, *Scientific American* devoted a special issue to the subject of intelligence.[29]

The figures presented in *The Bell Curve* are not disputed so much as are the explanations for them. Those who detect the influence of genetics are opposed by those who say the findings have more to do with social inequalities and test bias. Advocates of testing point out that the norms for modern tests are established on a mix of ethnic and social groups and have proven equally predictive of future achievement in all of them. Tests stripped of all vocabulary to eliminate cultural content have not pleased anybody because these tests predict less well for every group.[30] To visualise the statistics presented, one must picture separate bell curves for the racial groups, not fitting exactly one above the other because the averages are moved a little way apart. It will be obvious that there is still a lot more overlap than separation. In fact, one of the outstanding names in intelligence testing, Professor H.J. Eysenck, pointed out that, even with fifteen points of IQ separating the averages, if you tried to identify blacks and whites on the basis of IQ scores alone, you would be right only 5 per cent more often than if you decided by tossing a coin.[31] So although the hereditarians and the environmentalists argue vehemently about explanations, both sides are at pains to distance themselves from any taint of racism. An individual's ethnic origin tells us nothing about his or her intelligence. Everybody is agreed on that.

CHAPTER 13

GENES AND BEHAVIOUR

E VERYBODY ACCEPTS THAT BEHAVIOUR is predictable to some extent and that different kinds of animals feed, breed and relate to each other in characteristic fashion; we would be amazed to see a pigeon lifting its leg against a lamp post, or a dog going to roost on a wall for the night. These differences are innate and have evolved. Most people also know that very young puppies don't spray lamp posts and very young birds are not good at perching. These innate behaviours begin to show up only at a certain stage of development, like walking and talking in humans. It is also common knowledge that people of all groups behave differently at different ages, and that behaviour differs between individuals. Non-humans also differ individually and the differences can be more profound than most people probably realise. Variation is the stuff of evolution. I know I've said that before but it is worth emphasising.

On the other hand, it is also true that cultural and individual details of behaviour are often learned during an animal's lifetime and are not coded in its DNA. Behaviour that has no basis in DNA could not be passed on by heredity. So not all behaviour can evolve. What, then, is the relationship between genes, learning and behaviour? These are big questions for a small chapter, and will have to be tackled accordingly.

TEMPERAMENT AND PERSONALITY

We must always have known that human beings come in a wide range of personality types and, because we are social animals, we must always have been aware that some types are more likely than others to behave in particular ways. The everyday words with

185

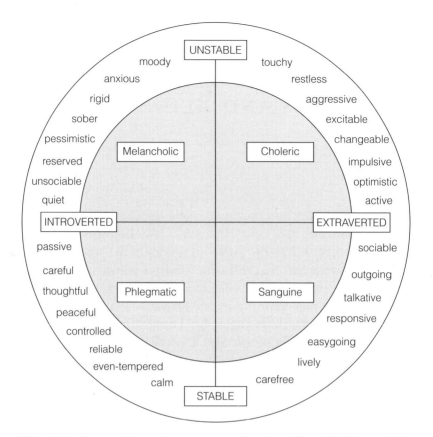

The circle of personality and temperament (Reprinted from H. Nyborg (ed.), *The Scientific Study of Human Nature*, 1997, p. 19, with permission from Elsevier Science)

which we describe human dispositions (shy, ambitious, sociable, quick-tempered, easygoing etc.) are based on interpretations of the behaviour that we see, and for thousands of years there have been attempts to make some systematic sense of it all.

Hippocrates, and later Galen, believed that the balance of the four humours (blood, black and yellow bile, and phlegm) could account for the state of both mind and body. As we noted in Chapter 1, an excess of one of these fluids was thought to make people (respectively) sanguine, melancholic, choleric or phlegmatic. Galen came up with nine temperamental types (Latin *temperare*, 'to mingle') based on different proportions of the humours.[1] More recently, psychologists have listed hundreds of trait names and tried to reduce them to a few supertraits. Nobody has been able to detect any one underlying factor, such as the *g* in intelligence, but there is some measure of agreement that all traits can be included within five supertraits. The big five are given different labels by different authors, but the following do make sense in ordinary language: extraversion, agreeableness, conscientiousness, neuroticism, intellect.[2]

The idea that traits can be at the opposite ends of a continuum has also been around for a long time. For example, the father of experimental psychology, Wilhelm Wundt (1832–1920), thought that with regard to feelings the melancholics and phlegmatics lay at the slow end of a speed-of-change continuum, with the sanguinists and cholerics at the fast end.[3] This slow–fast axis became the introvert–extravert dimension of the Swiss psychiatrist Carl Jung (1875–1920).

In London, Professor Hans Eysenck and his colleagues spent decades doing the most rigorous statistical analyses of personality traits and helped to establish the introversion–extraversion axis as one of the two major dimensions of human personality, the other being the neuroticism–stability dimension.[4] This was the weak–strong axis in Wundt's terminology, based on strength of change. Depicting the dimensions as compass points of a circle is a useful way of schematising the full spectrum of human personality types. We can all be located somewhere in the circle.

Now, we know from the study of twins that personality traits are heritable; that the variety of personalities in a human population is largely a consequence of genetics. So to some extent we can see how a range of behavioural types is part of the human evolutionary package. I am talking about individuals here; there is not much evidence to support the view that different *populations* of people breed true to any behavioural type, despite the fact that ethnic stereotypes have been popularised since Roman times. One example for which there is solid evidence concerns the behaviour of Asian and European babies. It was noticed 30 years ago in California that infants of Chinese descent were much more placid than those born to European parents. Subsequent studies on 4-month-old babies in Ireland and China also found that the Chinese infants were less fidgety and spent a lot less time crying and fretting.[5] It seems that this cannot be explained by differences in baby care; the two groups, on average, are temperamentally different. I am not aware of any follow-up studies to see how these infant temperaments relate to adult personality.

Eysenck and others have demonstrated that introverts tend to be more sensitive to stimulation, and are less tolerant of it, than are extraverts. Given the choice, extraverts set volume levels almost 20 decibels higher than do introverts.[6] The *arousal hypothesis* is the idea that introverts have a higher level of arousal in the brain to begin with. Consistent with the Chinese infant findings, it has been recorded that Asian–American psychiatric patients with anxiety symptoms need lower doses of the appropriate medication than do European–Americans with similar symptoms.[7]

Many people find the study of personality and individual differences quite fascinating, and of course there is no shortage of literature on the subject, but I suspect others will feel that this is really only the fringe of the genes/behaviour issue. Temperament, personality and aptitude obviously influence our choice of activities in life and the way in which we pursue them, but how are we to account for the core behaviour of a species? Why is it that nearly all individuals of a species, regardless of disposition, behave in a characteristic way? It is at this point that words like instinct and hard-wiring come to mind, but before we get into that I would like to set a behavioural scene based on an imaginary kind of bird. I hope you will bear with me.

ONCE UPON A TIME ...

I'll call the bird a furtive snatchbill and name it Phlik.

One day Phlik flew into a tree overlooking a stream. To his delight, but also with a stab of fear, he spotted scores of tiny, wriggling ooze-burglars stranded on a patch of damp sand. The little creatures were a delicacy, but the sand was in a very exposed place and this was stealth-hawk territory. A snatchbill, out in the open with its eyes to the ground, would be easy prey for a stealth-hawk. After a moment's hesitation, Phlik threw back his head and shouted 'fooood'. In no time at all, three more snatchbills appeared and the four of them went down to feed. It was better to share the tasty morsels and play it safe; four pairs of eyes, up and down at different times, are more efficient than one pair in looking out for danger.

Phlik flew on down the stream and arrived at a stretch that he patrolled daily in search of things washed out of the shallows. There were no trees close by so he wasn't afraid of stealth attacks.

Another snatchbill was already there; it was one that Phlik had chased off the day before. This time, Phlik noticed the incessant movement of prod-hoppers at the waterline so he allowed the other snatchbill to stay. On a good day like this there would be enough hoppers for two, and a couple of birds could more quickly chase off other gate-crashing snatchbills if need be. Besides, you had to get prod-hoppers while you could; they didn't hang about while you were off chasing a rival.

Meanwhile, back at the nest, a snatchbill drama was being played out. Phlik's mate, Phyllis, was off her eggs, dragging one wing and being pursued by a lolloping, flop-eared animal with a collar round its neck. Just when it seemed that the brute would catch her, Phyllis put on a sudden spurt and got ahead again. She did this several times until, finally, she flew into the air and began to circle the bewildered beast, scolding and squawking until she made it turn for home with its tail between its legs. Feeling that the danger was past, Phyllis sailed back to the nest and got there not a moment too soon. A cow was blundering about near the eggs, and if Phyllis hadn't been there to be noticed the cow might well have trodden on them. She settled down to incubate again and await the return of her mate. It had been her very first performance of the distraction trick.

The eggs were quite close to hatching and two were already pipped. That is to say, the chicks inside had broken through the shell at one point and would soon be treading a circuit inside, hammering out a neat ring of perforations. The line of fractures would allow each chick to push the shell apart, like heaving the lid off a pillbox. Phyllis sat tight and answered the chicks' muffled cries with special, quiet calls.

Phlik was nearly home now and had finished his foraging on a good note. He had spotted a juicy mud-slob, larval stage of the bowel-fly, lying by a rock. There had been no announcing his find this time; a single mud-slob couldn't be shared, so Phlik had kept quiet about it and was now carrying the creature back to his faithful Phyllis.

Well, almost faithful. In truth, one of the eggs in the nest wasn't Phlik's at all. There had been another male snatchbill with an exceptionally long tail which Phyllis had found quite irresistible. She wasn't to know that it was a false tail, glued on by a cunning zoologist-person. Phlik never suspected he had been cuckolded and of course Phyllis didn't tell him; she had no way of talking, or even thinking, about such things. Furtive snatchbills are just birds, with a brain not much bigger than an aspirin.

Utter nonsense? Not entirely. All the creatures are fictitious but the bird behaviour is based on serious, quantified studies of different kinds of birds. I rolled them into a fable in order to cover a lot of ground without getting into serious and more lengthy ornithology. The activities of Phlik and Phyllis do illustrate the kind of survival behaviour that scientists since the 1960s have been trying to explain. There are important issues involved and not just for birds. Some recent insight into human behaviour was reached largely via the understanding gained from research on other species.

Under the headings that follow, in a stepping-stones sort of way, we can pick out the main stages in the path to our present understanding of behaviour—including that of humans.

DOGS, CATS, PIGEONS AND UTOPIA

Compared with human society, the lives of Phlik and Phyllis seem very simple, but compared with the life of a limpet, a bird's affairs are immensely complicated. There is both continuity and overlap in animal behaviour as in other aspects of life. At the simplest end, the primitive mechanisms which enable limpets to respond to a stimulus are still critically important in birds and mammals. Sneezes, flinches, blinks and the squirting and flowing of bodily fluids at appropriate times still have their place in vertebrate behaviour, even though we do so much that goes beyond that level.

Ivan Pavlov (1849–1936), the famous Russian physiologist, won the Nobel Prize for his work on digestive secretions in 1904. He also discovered that reflex actions could be made to occur in response to a stimulus that wouldn't normally produce them. Dogs could be conditioned to slaver at the sound of a food bell, even without food. A puff of air on our eyes will make us blink, but if the puff is preceded by a noise then after a few trials we shall find ourselves blinking at the noise. The time interval between the two stimuli (noise and air in this case) is important: about half a second is usually most effective in humans.[8] This sort of associative learning, known as classical conditioning, has its uses in laboratory research but does very little to explain complex behaviour. It cannot produce new behaviour; it can only find new triggers for old responses—although the new triggers can include all the emotion-rousing devices of the arts. Conditioning in nature presumably has survival value in helping animals to associate smells and colours with distasteful plants, and to establish other danger signals. Phlik became fearful at the sight of oozeburglars on sand because of past association with a hawk in similar circumstances. In humans, accidental conditioning is not always useful. Some time ago a dentist played what he thought would be distractingly merry music while he hurt me, and I've had an unpleasant reaction to the tune ever since. Instead of being distracted from the pain I became conditioned to the music.

An American psychologist, Edward Thorndike (1874–1949), studied a different kind of learning in cats and believed it could explain a lot more about behaviour. When a cat was locked in a cage it would become very active and give every appearance of trying to get out. By chance, a cat could push against the door latch and escape. When this happened, Thorndike would return the cat to its cage and record the time it needed to make a second escape—and a third and fourth and so on. What he ended up with, plotted on his clipboard, was something that looked like a learning curve as the cat got steadily faster at opening the door. Thorndike called it trial-and-error learning and assumed that it must be accompanied by some sort of neural process, but he argued that there was no need to understand the neural changes in order to explain the learning. For practical purposes we could just say that the behaviour was 'stamped in' because it was immediately followed by the reward of the open door. No insight was needed for this to happen and none was assumed.

Thorndike's important contribution to psychology was to show that behaviour could be studied and measured without knowing, or even needing to know, anything about the mental processes that might accompany it. It took a number of years for this approach to be applied in a major way to humans, but then *behaviourism*, as it came to be called, took psychology by storm and its influence is still with us. A professor of psychology at Johns Hopkins University, John B. Watson (1878–1958), is generally regarded as the founder of the movement, for it was his many publications that gave behaviourism its high profile in

America during the 1920s. The philosopher John Locke would have loved Watson. 'Give me a dozen healthy infants, well-formed, and my own specified world to bring them up in', Watson wrote, 'and I'll guarantee to take any one at random and train him to become any type of specialist I might select—doctor, lawyer, artist . . .'.[9] The blank-slate conviction must have been strong indeed.

But despite Watson's reputation as the pioneer, it was a Harvard professor, B.F. Skinner, who became the more widely known behaviourist. Burrhus Frederic Skinner (1904–1990) had no time at all for speculation about what might be going on inside the brain. That was the work of neuroscientists, and Skinner saw no useful purpose in describing states of mind with words such as fear, love, hatred, anger, enthusiasm or determination. In his view, these were nothing but names for states that had to be inferred from behaviour that could be seen anyway. So why bother with them? Why not stay with the observable behaviour? Behaviour was something tangible; it could be demonstrated, measured and controlled. Skinner didn't deny that inner states might exist, but in his view they were 'not relevant in a functional analysis'.[10]

Skinner would have described the behaviour of Phlik and Phyllis largely in terms of movements and activity patterns, which had been shaped by the same process that had built up the escaping skills in Thorndike's cats. Every element of behaviour that produces a rewarding result gets *reinforced* until it becomes part of the animal's established repertoire. Non-rewarding or detrimental behaviours tend not to be repeated. Obviously, the behaviour of different species is reinforced by different rewards; it might be food, drink, escape from danger, a mother's smile, a pay rise or just a good feeling. In Skinner's model, the principle is the same for all animals. Birds learn to find food, children learn to talk and write, and pilots learn to fly aeroplanes in exactly the same way. Skinner called the process *operant conditioning*.[11]

Skinner saw no value in thinking about the role of innate differences. He accepted that individuals were different and recognised that there was 'genetic discontinuity' between organisms so that some behaviour was specific to a species, but he considered such 'genetic units' to be rare, at least in vertebrates. In Skinner's words: 'The behaviour with which we are usually concerned, from either a theoretical or practical point of view, is continuously modified from a basic material which is largely undifferentiated'.[12] As we shall see, there is now good reason to believe that Skinner was wrong about this.

Professor Skinner wasn't shy when it came to promoting his ideas for more effectively controlling human society, nor had he any qualms about straying into the realms of ethics and social policy; one of his books was about an imaginary Utopia created by operant conditioning.[13] Yet for all the scorn of his critics Skinner cannot be dismissed as a mere armchair theorist. He demonstrated his operant conditioning on a variety of animals by training them to carry out highly complex behaviours in novel situations. Animals are most inclined to make the sort of movements that come naturally to them, and there must be limits to a repertoire, but by carefully timing the reinforcement of each little movement that would contribute to the complete performance, Skinner could even get two pigeons to play a kind of table tennis.

To my mind, anybody who can teach pigeons to play ping-pong must know something more useful about animal learning, and it is a fact that Skinner's methods are now standard practice among animal trainers. Skinner's name has also been immortalised in connection with the various puzzle-cages in which laboratory animals must press levers, operate light switches or otherwise demonstrate their yearnings and learnings. The cages are invariably known as Skinner boxes. Skinner's conditioning principles are also applied in computer programs that provide interactive tuition for people, and we can all think up thousands of examples of everyday human behaviours, from personal grooming to social conduct, which have been shaped by what Skinner described as the 'contingencies of reinforcement'. But research has moved forward on other fronts since Skinner's revelations, and nobody now believes that all the behaviour of any animal can be explained solely in terms of its experiences in life. Whatever philosophers may have thought about minds starting off as clean slates, environmental determinism doesn't fare any better than genetic determinism in the light of modern biology.

RUNNING ON INSTINCT

The notion of instinct has its roots in theology, not science. In the belief that only humans had souls, Descartes was committed to a position which denied all other species any capacity for reasoning. For centuries, while that view of life prevailed, the activities of Phlik, Phyllis and the dog would have been explained as nothing more than the visible workings of 'clockwork' brains. Humans lived by reason; all other animals lived by innate programs of behaviour which came to be called instincts. Those animals that could obviously learn, such as dogs, had to do so mechanistically, because any kind of reasoning or insight would require a mind, or soul, and souls were held to be uniquely human.

In the 1920s, after early experiments and observations on chimps, it began to look as though some non-humans could have flashes of insight after all. Work at a primate research centre on the Canary Islands produced the now classic descriptions of chimps scratching their furrowed brows then suddenly joining two sticks in order to reach a banana outside the cage, or piling two boxes to make a high step when food was suspended from the ceiling. The director of the research centre and author of these descriptions, Wolfgang Kohler, might have read a bit too much human body language into his chimps' behaviour, but he succeeded in making scientists revise their ideas about learning processes. Nevertheless, demonstrations of insight in chimps did not get rid of the distinction between learned behaviour and the hard-wired sort. It was still possible to think in terms of a clear divide between the two, and according to one authority: 'The chant among social scientists in the 1920s was that humans had no instincts'.[14]

Darwin devoted a chapter of *The Origin* to instinct, but his main concern was to show how instincts could have evolved, rather than to explain how they worked. It is easy to think of examples of behaviour that *must* be instinctive. Spiders weave their webs, ants tend their fungus gardens and chicks methodically break out of their eggshells with absolutely no prior experience; they evidently inherit the complete behavioural program in genetic code. But other examples are not so cut-and-dried: some 'instincts' wax and wane with the seasons, while others produce performances that improve with practice.

Not until the middle of the twentieth century did the study of instinct become scientific. The ethologists (three of whom were introduced in Chapter 10) were not concerned with teaching animals to do tricks; they studied the natural behaviour of animals with due regard to its evolutionary significance. In his *Study of Instinct*,[15] Niko Tinbergen, then at Oxford University, showed how animals could inherit behavioural responses to very specific stimuli. In little fish called sticklebacks, the red breast of a breeding male acts as a sign stimulus for a rival male to attack. The rest of the fish is not taken into account: a red breast on a vaguely fish-shaped model will do. Similarly, the swollen belly of a female fish that is carrying eggs is the cue for the male to start courtship. A chain of stimulus–response behaviours (fixed action patterns) then culminates in the female discharging her eggs in a nest and the male pouring his sperm over them.

Tinbergen also showed that the stimuli that occur in nature could be made more effective. Up to a point, bigger bellies and bigger eggs for incubating birds can be more effective stimuli than those which nature provides—as was the longer tail on the male bird that seduced Phyllis in my story. It is as though natural selection has worked through all the pros and cons and arrived at the best compromise for the stimulus, but built a little in reserve on the demand side. If girlie magazines are any indication, many western men are attracted by more prominent breasts than are usual on women, but I don't know whether this has been studied scientifically.

A human instinct that has been the subject of scientific study is the smiling response in infants. The main trigger is a pair of eyes. For 6-week-old infants a picture of two conspicuous eyes on a blob of a face is more effective than a realistic face. The eyes can even be one above the other (as they would be in a real face tilted sideways) instead of side by side. As infants mature, images that are more realistic become more effective. No learning is necessary for the smiling response to develop; blind babies do smile.[16]

In adults the same facial expressions of anger, fear, disgust and amusement are recognised across cultures and probably have a strong innate component, even though they can be modified by learning. In facial expressions, as in activities such as walking, preening or copulating, the basic movements come naturally, having been coded and transmitted in DNA, but there is still scope for the performance to be altered by experience. The dive response of newborn babies, in which they hold their breath and swim under water, seems to be a fixed action pattern stimulated by immersion. Human babies, dangled with their feet just touching the ground, will make walking movements even though their legs are not strong enough to support them.

Darwin could see for himself that instinctive behaviour often had a certain plasticity. He described instinctive actions as those which are typically performed in the same way by animals of a species, without experience and without knowing the purpose of their actions, but he then went on to say: 'I could show that none of these characters of instinct are universal. A little dose ... of judgement or reason often comes into play, even in animals very low in the scale of nature'.[17]

Science has now confirmed that much complex behaviour is no longer to be regarded as either instinctive or learned, but rather as instinctive *and* learned. It has become apparent, as it did with the nature–nurture issue, that learning and innate programs interact, and one does not exclude the other.

INSTINCTS AND OPPORTUNITIES

We don't need science to show us that an individual's behaviour often depends on its stage of maturity. Babies babble before they talk, crawl before they walk and so on. But another of the ethologists who shared the Nobel Prize in 1973 demonstrated how some patterns of behaviour become fixed at certain critical times in an animal's life. You may have seen published photos of Konrad Lorenz as a white-haired figure leading a line of devoted goslings who thought he was their mother. Geese and other birds direct their social behaviour towards whatever moving object takes the parent role. The chicks become *imprinted* on the sight and sounds of the mother figure during a sensitive period shortly after hatching. Birds hatched in incubators and hand-reared will become imprinted on humans and fail to develop normal responses to their own species—even to the point of trying to mate with the keeper's head in the case of hand-reared aviary birds.

Birdsong can be a delicate mixture of innate and learned behaviour. Chickens and some other species will start to produce normal calls even if they have never heard another bird, but studies of some songbirds have shown that they always sound like beginners unless they are allowed to hear a proper example from an older bird at the appropriate stage of their development. Even then, they need to practise. Birds experimentally deafened, shortly after hearing the example, still cannot perfect their songs, and birds deafened at birth do even worse than those with normal hearing which have been raised in isolation. It appears that there is a genetic template that gets the bird started and directs it to learn the correct song—but timing and opportunity can be critically important.[18]

The distraction trick, which Phyllis turned on to lead the dog away from her nest, is performed by various plovers and similar birds around the world, and it must be very comprehensively programmed in DNA because a bungled first attempt could be fatal. There may be scope for improving the performance, but from the start the bird has to decide which animals to lead away and which (like the cow) to stand up to. It must also decide how much risk to take and what distance to cover.

Brains can evidently evolve to provide for finely programmed behaviour at one extreme and open learning at the other. Between these extremes there is the possibility of learning with some degree of genetic influence, and this seems to be the way that a large proportion of vertebrate behaviour is acquired. Recalling that brains are metabolically expensive, this arrangement makes evolutionary sense. Ant societies, with their workers, gardeners, soldiers and slaves, show the organisation that can be achieved by animals with pinprick brains living in a micro-environment where there is no need for thought and judgement. In the more complicated world of vertebrates, flexibility of behaviour has been acquired at great cost in terms of brain energy, but natural selection will not have passed on any wastage in this regard. Because energy is life's common currency, advantage lies with those who need less of it, other things being equal. Innate devices such as fixed action patterns, learning templates and limited learning periods, whatever these are in neural terms, have proven their worth as reliable, ready-written working modules that can be passed on for elaboration by more flexible circuitry as necessary.

In a similar way, natural selection appears to have fixed those behaviours which make for efficient foraging. Sparrows, when they find crumbs that can be shared, tend to advertise their find by chirping, but they tend to keep quiet if there is an indivisible lump—as did Phlik with his mud-slob. And, like Phlik with the hoppers, wagtails patrolling the waterside tend to tolerate a rival when the pickings are rich enough for both.[19] Nobody is suggesting that the birds can work all this out for themselves; instead, natural selection has filtered out the most successful strategies wherever genes have been involved, and the birds' brains now work accordingly. With humans, flexibility has gone further than in any other species, but there is plenty of evidence that genes, contrary to earlier belief, have a powerful influence on what we learn and when we learn it, and not only at the level of reflexes and fixed action patterns.

MODULES OF THE MIND

Frederic Skinner was so convinced that nearly all learning had a common foundation that he wrote a book to explain how people learned to talk in the same way that they learned most other skills—by the principles of operant conditioning.[20] In 1959 Skinner's book was reviewed by the linguistics guru Noam Chomsky, of the Massachusetts Institute of Technology.[21] Chomsky's 30-page critique was a demolition job, and it now marks the point at which mainstream psychologists gave up trying to explain language in behaviourists' terms. Language was different. 'The fact that all normal children acquire essentially comparable grammars of great complexity with remarkable rapidity', wrote Chomsky, 'suggests that human beings are somehow specially designed to do this'. He then went on to point out that there was nothing mysterious in that, because 'tendencies to learn in specific ways' had been carefully studied in lower organisms. The point about grammar is crucial to the argument. Not only has the average 6-year-old accumulated a vocabulary of perhaps 13 000 distinct words of the sort that can all be arbitrarily paired, but even infants of eighteen months know how to arrange a verb and noun to make meaning. In every language of the world babies are saying the same sorts of things at a similar age and are learning new words almost by the hour. The brains of human infants have apparently evolved mechanisms for speech learning just as those of songbirds have been selected for song learning and that of Clark's nutcracker has acquired what it takes to memorise thousands of little hiding places.

Children do need to practise and to hear themselves, as do

songbirds, and of course there are sensitive periods. Children can learn two or three languages at the same time and end up sounding like a native speaker of all of them, whereas adults can struggle for years with one foreign language and never be able to pass it off as the mother tongue. Babies under six months old can distinguish the shortest speech sounds (phonemes) of foreign languages, while adults can't tell them apart even after 500 trials. By ten months, the same infants can distinguish only the phonemes of the language they are learning. When *Homo* evolved speech, the pressure was on learning the language of the group— and quickly. The need to learn a foreign language, like the need to read and write, did not exist.

The modular view of language learning has no serious challengers today, as many will understand from a popular book by another professor at MIT, Steven Pinker of the Department of Brain and Cognitive Sciences. Pinker's book, *The Language Instinct*, which was published in 1994 to well-deserved acclaim, ranges over the whole subject of language, from the real meaning of deep structure to such delightful trivia as 'sex between parked cars'.[22] It was from Pinker's book that I took the details for the previous two paragraphs.

We should not get carried away with modular models of the brain, but there may well be other specific learning mechanisms in human beings. One authority provides a bookful of evidence for the existence of mathematical 'starter kits' probably located in a little knot of brain tissue located above the left ear.[23] For what the observation is worth, this is the same region (the inferior parietal lobe) which was recently found to be enlarged in Einstein's pickled brain.[24]

Very young babies also appear to have some understanding that living and non-living things have different properties. In experimental arrangements, they show no great interest if a moving person stops and another starts moving, but if the same thing happens with billiard balls there seems to be an expectation that one must hit the other to start it moving.[25] Pinker observes that people everywhere, without education, classify living things with a different logical structure from the one they use for other objects; a snake is always a kind of animal whatever we do with it, but a piece of old metal might be a toy, a tool or a container. The names we give to the living and the non-living may even be stored in different parts of the brain—but more about brains in Chapter 15.

TIME SCALES

Special intuitions regarding plants and animals would, of course, make perfect sense in the context of our ancestral environment. It was Tinbergen who pointed out that to make sense of behaviour in a natural setting we must look beyond the immediate function of the behaviour. It isn't enough to see the obvious purpose in terms of finding food, getting a mate or whatever, we must also consider the behaviour in the context of the animal's evolutionary history. We have already acknowledged the risk of coming up with 'Just So' stories in trying to account for every characteristic in terms of evolutionary adaptations. It is a risk that may sometimes be worth taking, but it may be possible to collect supporting evidence.

Think of those female birds that prefer males with long tails. Why should this be? Where's the survival value in having a tail that is more of a hindrance than a help in flying? Where is the evolutionary fitness in that? One possibility is that the long tail has become a signal that says the bearer must be something rather special in *other* respects because only exceptional individuals could survive with such a handicap! When the handicap principle was first put forward in 1975 there were many sceptics, but we now know from experiments and observations that male barn swallows with longer tails do tend to have more heritable parasite resistance and they are favoured by females—even to the extent of extra-pair copulations. It is the parasite resistance that improves survivorship in the brood.[26] The female, of course, has no idea why she prefers long tails. Genetic recipes merely get selected; they don't come with explanations.

There was a time when those with an interest in natural history were breaking new ground simply by finding out what creatures did in the wild. Asking *why* they do it has opened up new horizons in science. Behavioural ecologists these days are particularly interested in examples of behaviour which, at first sight, appear to be at odds with evolutionary theory. In short, they are keen to explain the problems of evolutionary theory rather than to demonstrate the predictable. To do this it is essential to look beyond the immediate, or proximal, cause of the behaviour and to think about the ultimate cause in evolutionary time. The part that heredity plays in behaviour must have been fixed by evolution over many generations, whereas details of the same behaviour, which have been learned, could be changed with a new experience.

This last point may seem patently obvious, but arguments based on jumbled time scales still account for an awful measure of confusion. When we talk about humans in terms of behavioural ecology (or sociobiology or evolutionary psychology for subject-splitters), we invite the input of some influential academics who are evidently not accustomed to thinking in evolutionary terms at all. Until quite recently, social scientists have generally been content to follow the doctrine that was established during the 1920s, when it was popular to deny the very existence of human instincts. In the Standard Social Science Model (SSSM)[27] as this mindset has come to be called, humans are controlled by learning and culture, whereas all other animals are controlled by biology. And human learning is of the general-purpose type, in which we all start off with a spotless mental slate. For a long time, any departure from the SSSM was not only incorrect, it was improper. Genetics was a dirty word.

Well, the SSSM is breaking down along with some of the old academic demarcation lines, but entrenched ways of thinking can be persistent. In one recent textbook, a leading British sociologist defined an instinct as a '*complex* pattern of behaviour that is genetically determined', and he felt able to dispose of the topic with the single sentence: 'Most biologists and sociologists agree that human beings do not have any instincts'.[28] That was in 1997. The fuss about sociobiology is the subject of the next chapter.

CHAPTER 14

THE SOCIAL DIMENSION

A NUMBER OF WRITERS have remarked, as I did at the end of Chapter 10, that the hostile response to Edward Wilson's *Sociobiology* was surprising. In retrospect, we can see some explanations for it. For a start, by bringing humans into his thesis, Wilson clearly caught the attention of many people who would not otherwise have given the book a second glance. Inevitably, among the new recruits to evolutionary thinking, there were some who expressed strong views while they were unfamiliar with the subject and unaccustomed to the trade language of evolutionary biologists. On the face of it, here was an ant specialist writing about highly sensitive social issues such as sexual behaviour and family relations, furthermore provoking the sociologists by telling them that their subject was stuck in a rut and they were no further ahead than pre-Darwinian naturalists had been—strong on description but lacking any theoretical framework for explaining what they saw. How dare he!

Some writers who joined the sociobiology bandwagon went much further than Wilson had done in trying to explain human behaviour in terms of evolutionary adaptations. There was deliberate sensationalism as well as careless writing and overenthusiasm in passing off 'Just So' stories as accepted explanations. This provided welcome ammunition for the critics. Not surprisingly, they made full use of Dawkins' paragraph about animals being 'lumbering robots' through which genes manipulated the outside world.[1] Another favourite was the claim by Michael Ghiselin, a biologist at the California Academy of Sciences, that all altruists must really be hypocrites.[2] In an overreaction, the critics in turn began to demand impossibly stringent proof that *any* human behaviour was an evolutionary adaptation. Too much mud got slung and, as two analysts recently observed: 'it's hard to construct good theory while mud wrestling'.[3]

In this chapter I want to point out the chief sources of misunderstanding, outline some of the main issues that sociobiology brought into the spotlight, and have a look at the current position.

THE FAILURE TO COMMUNICATE

If biologists say that they are of the opinion that birds' ancestors first grew feathers in order to keep warm, rather than to fly, other biologists will understand the evolutionary shorthand. They will understand it to mean that feathers first arose as insulators rather than as an aid to flight. They will not read into this any suggestion that birds' ancestors *wanted* to grow feathers, *tried* to grow feathers, or *understood* why they were growing them. The switch from scales to feathers must have had a selective advantage under certain circumstances, and birds' ancestors were the unwitting phenotypes in that particular numbers game.[4] Evolutionary literature is full of this kind of shorthand, and it can mislead readers who don't understand how natural selection works.

When mechanics talk of slave and master cylinders we all understand the relationship between the components involved. Similarly, biologists often use analogies from human society for labelling social arrangements in other species. On the topic of insects they might talk of castes, slaves, armies, workers, queens, royal attendants and so on. Other animals have spouses, suitors, harems, bullies, tyrants and cheats. These are convenient and welcome terms in a subject notorious for its Graeco-Latin vocabulary, and they are commonly used without implying exact social parallels between humans and other species. All the same, some social scientists are upset by this and detect serious political implications. By using false metaphors, they claim, sociobiologists can help to justify bad behaviour in humans.[5] Feminists in the 1970s protested passionately about the use of the word rape in descriptions of non-human sexual acts, even though the females of many species are often forced into sexual submission by stronger males.[6]

And then there is the language of economics. In this case there are valid parallels to be drawn. Despite there being no conscious trading of commodities among non-humans (though this can't be ruled out for chimps), animals do need access to resources and often have to work hard for a living. They do make investments of time and energy and receive benefits, pay costs or lose their stakes. In studies of time and energy we can apply the language of economics to a bird colony no less than to a community of hunter–gatherers, but the politically vigilant are always ready to pounce. A common claim is that sociobiologists talk the language of capitalist exploiters and try to make it sound natural. One critic particularly quick off the mark in response to Wilson's book was Professor Marshall Sahlins, an anthropologist at the University of Chicago, who claimed that social Darwinism had returned to biology as 'genetic capitalism'.[7]

And so it goes on; overreaching claims, politics, ideology and mischievous journalism are still being stewed with some dumplings of pure misunderstanding. One misunderstanding, in particular, is such a nuisance that it should be cleared up here and now.

TIME SCALES AND CONSCIOUSNESS

The *ultimate* cause of behaviour, the reason for its being selected in evolutionary time, should not be conflated with the *immediate* cause—which is what the individual knows about in the present. For example, men from all cultures want young women for mates in preference to aged ones. This is true regardless of whether the men want to have children by the union. Younger women are sexually more attractive to men, and this attraction is the immediate cause of the man's behaviour towards her. But viewed in evolutionary time those feelings of attraction, the whole emotional bundle, are part of the genetic recipe that was winnowed out by natural selection because it resulted in the genes for that recipe being passed on most efficiently. A man whose passions were aroused by grey hair and wrinkled skin might have been well satisfied with life but any genetic basis for his feelings wouldn't have had much future. We choose our partner for an immediate reason and we needn't consider, or even be aware of, the more distant cause of our behaviour.

To the extent that our reproductive and child-rearing behaviour is genetically influenced, the basic human pattern must have been shaped by natural selection and, all things considered, must represent the most efficient human breeding package that has been available for selection to work on. I therefore find it impossible to agree with the psychologist who, in attempting to dispense with sociobiology in a couple of pages, questioned the existence of maternal or other reproductive instincts on the grounds that 'the advent of birth control renders this dubious line of reasoning meaningless'.[8] This is like saying that because we never climb trees in our city clothes it is meaningless to consider the adaptations that we acquired during that phase of our evolution. Today's choices by individuals, about breeding or anything else, cannot wipe out our evolutionary past—though they may well affect our evolutionary future. The reference to birth control implies that we have overcome our evolutionary legacy by discovering contraception because now we can enjoy sexual frolics without wanting to breed. In fact, that is the way it always was. Until very recently, no animal on earth knew the consequence of its mating.[9] Rutting stags, strutting peacocks and their choosy mates are not driven by the desire for fawns or chicks any more than the mating mayfly is thinking about nymphs. For them, there is only the immediate sexual goal. With contraception, and especially without love, our goal is much the same.

The objection to the word rape for non-humans seems to be based on a similar misunderstanding. Human rape, it is said, should be seen as an act of violence, without any reproductive intentions. Absolutely. The same is true of non-human rapists: they are not consciously trying to reproduce either.

There is no reason to doubt that our common reproductive behaviour has evolved by natural selection. It is obviously adaptive; its genetic recipe has proven to be a winning formula and is currently producing some 130 million copies of itself in live births every year. But what about all the variations that exist? What about the finer points of it all? What about monogamy, polygyny, loyalty, jealousy, infidelity, and all the different attitudes to parenting around the world? Speculations about these issues, spelled out in terms of faithful wives, promiscuous husbands, wicked stepfathers, selfish siblings and the like, are those which grab the journalists' attention. There are still plenty of writers who insist that all this variation is entirely cultural and that genes don't come into it. But that's what they used to say about language. We will stay with the reproductive theme for one more section.

THEMES AND VARIATIONS

It is a fact of life that the male of many species releases teeming millions of copies of his genotype in sperm, whereas females with eggs that develop internally produce only a few. Moreover, in species like humans, the female has to spend a significant part of her life in pregnancy and nursing if she is to raise even a tiny number of offspring successfully. As a consequence, at the end of reproductive life, a woman may have mothered and reared a dozen children but a man can have fathered a thousand. Equally obviously, if the sexes exist in roughly equal numbers, not many men can father hundreds of children unless many others father none at all. In a colony of elephant seals, that is the normal state of affairs: 4 per cent of the males father 85 per cent of the offspring, and for as long as they can keep fighting off the challengers the few dominant males will be passing on lots of competitive genotypes. Future champions will do exactly the same.

In human societies, a few high-ranking males have been known to monopolise a large proportion of the local women, but polygyny on that scale is hard to maintain if only because deeply frustrated young men can be far more treacherous than any number of hopeful seals. With or without a recognised system of polygyny, it remains a fact that women *cannot* have children in numbers very far from the average whereas it is possible for males to deceive, charm, bully or bribe their way into becoming fathers time and again. The numbers add up, so one might expect those men among our ancestors who took the risks and got away with it to have left the most offspring. Why, then, has the genetic recipe for male philandering not come to outnumber others in the way that the recipe for dominant bulls has come together in elephant seals? Or has it? Anyone who doubts that males are the risk-takers in our species need only look at the statistics for criminals: men outnumber women by about nine to one.[10] There is also plenty of evidence that men, on average, are more promiscuous than women. In one study, using bogus propositioners, 75 per cent of men accepted the offer of sexual favours from total strangers, but of the women who were approached there were no takers.[11]

It is easy to show with a computer that the genetic make-up of the males who fertilise the most females, other things being equal, will come to dominate the populations in which they live. The crucial phrase here, as in so many arguments, is *other things being equal*. Obviously, other things will not be equal in a real society of primates, and especially in humans. Impregnating more females does not necessarily get more infants through to maturity to carry on the genetic line. Without adequate parenting, the outcome of 'illicit' fertilisations in our ancestral societies might have included a lot more abandoned babies, malnourished infants and higher child mortality. We can only guess at the way the numbers might have played out through selection. Similarly, daring attempts to impregnate more females would have resulted in disproportionate losses among the ranks of the risk-takers, but we cannot know *how* disproportionate. And we still haven't considered the selective forces that might have been at work among the females of our species. In the full complexity of behavioural evolution, what comes out in the end is what we see: some obvious characteristics and tendencies, but not much that is hard and fast among the many variables.

Parental investment is the energy input that goes into rearing offspring. In most vertebrates, though not all, the female puts much more time and energy into parenting than does the male. Any behaviour that helps the mother to get her offspring to maturity should help to carry the mother's genotype into the next generation. Women who repeatedly got themselves pregnant by men who promptly deserted them would not be the ones likely to leave most grandchildren. On the contrary, the forces of natural selection would be likely to favour women who were most skilled at detecting and retaining fertile but also supportive mates for the many years that human child-rearing demands. So what do we find in this regard?

Like men, women are sexually active more or less continually; they don't come on heat with the seasons. Unlike apes, which also have a menstrual cycle, women remain sexually receptive outside the time of ovulation, and there is no physical change to indicate when that time occurs. Not even the woman herself knows exactly when in the month she is most fertile, so in terms of passing on behavioural recipes the continuous pair bond works very well. And there is evidence that much more subtle characteristics of women could be adaptations for establishing and maintaining these lasting relationships.

It follows that women would be favoured, in evolutionary terms, if they were able to avoid being left to rear babies as single parents. We should therefore not be surprised to find that, on average, women prove to be better than men at detecting cheats and liars. The famous 'feminine intuition' of novels may be well founded. It also follows that women would be less harmed (again in evolutionary terms) by occasional male infidelity than by desertion. Female infidelity and deceit, on the other hand, can result in the deceived male helping to rear a rival's child and so helping to launch a competing genotype into the next generation. Males who were genetically disposed to be cooperative cuckolds would be weeded out of the population. Mate-guarding and resistance to being cuckolded would be favoured. The prediction is that men should be likely to get more emotional about sexual deception, whereas women should be more distressed by the threat of desertion than by the act of sexual infidelity. Again, such evidence as there is supports the prediction. Men can become murderous on discovering a wife's adultery, but to women it more often matters whether a man is being unfaithful just for thrills on the night or there is a transfer of affection—an 'affair'. This is reflected in the laws of some countries as well as in the crime figures. As always, individual exceptions abound but the overall picture around the world does seem to fit the evolutionary interpretation, which pleases some people and infuriates others.[12]

There are conflicting research findings on the incidence of child abuse by human step-parents.[13] In many animals, selection has favoured the genotypes of those males which kill their stepchildren when they take a new mate because, statistically, this has improved the chances of their fathering more young of their own. Lions have inherited that behaviour and so have langur monkeys. These animals cannot know why they feel the way they do, but perhaps Moses thought he could rationalise his feelings when, according to the Bible, he instructed his military commanders to 'kill every boy and kill every woman who has had sexual intercourse, but keep alive for yourselves all the girls and all the women who are virgins'.[14] That passage is one the sociobiologists are fond of quoting!

THE BIG TABOO

Among the behavioural traits found in virtually every society is that of avoiding incest. Only in graffiti is incest a game for the whole family to play.

Inbreeding is medically hazardous, a proven reason being that related individuals have enough genes in common to raise the chances of harmful recessives coming together in a mating. In a range of mammals where siblings have been mated in captivity, the average mortality rate among the offspring is about 33 per cent higher than in unrelated matings, and of course there can be innumerable problems that are less than lethal.[15] Where such a strong selective disadvantage has been at work we must expect to find that siblings are less than eager to mate with each other, or with their parents. Sure enough, among those bird and mammal species which live in families, nearly every study has found incest to be statistically rare.[16]

In behavioural studies of family-living mammals and birds it has been found that sexual attraction seems not to develop between family members or those raised as family members. This is apparently the incest-avoidance mechanism that has evolved in these species. After individuals have been reared together from infancy, regardless of their biological relationship, they don't usually pair up sexually. Wealthy families in Taiwan had a tradition of adopting young girls as future daughters-in-law and rearing them together with their future husbands. Out of nineteen such adoptions in one village, seventeen couples refused to consummate their marriages. In Israeli kibbutzim, unrelated children are reared communally in nursery groups and daycare centres as if they were all in one big family. A study of 2769 marriages between second-generation kibbutz adults found that only six marriages were between members of a peer group of any one kibbutz.[17]

In societies of all kinds there are cultural prohibitions, backed by laws and penalties for preventing incest. On the face of it this looks like nothing more than a common cultural pattern, but with a biological perspective it appears far more likely that human cultures are adding to the enforcement of a good behaviour bond written in DNA.

BLOOD IS THICKER ...

Kinship and the evolutionary problem of altruism came up in Chapter 6, and it is time to revisit it in the human context. I suppose I must also bring back the word 'fitness' because it is probably impossible to turn to any further reading about altruism that doesn't rely on a proper understanding of fitness in the evolutionary sense. Fitness is a measure of the reproductive success of an individual (or whatever is replicating) relative to that of another, and it only makes sense to think in terms of survival and reproduction in subsequent generations, not just the current production of offspring. Successful traits are not predictable; we can expect an outcome but be completely wrong. We see this constantly when introduced animals get out of control or when our careful wildlife introductions run to extinction.

We saw that the nub of the altruism problem is how to explain the evolution of behaviour that works against the fitness of the altruist to the advantage of the one being helped. The solution put forward by William Hamilton, with its later refinements and elaborations, explained most of what would otherwise have remained a puzzle. It showed, mathematically, how an animal can pass on more of its genes by dying to save relatives than by living while they perish. This concept of individual fitness plus the accumulated consequences of kin selection is known as *inclusive fitness*. As long as the act of helping relatives, with or without loss of life, serves to promote one's own genes, natural selection will automatically favour recipes for helping.[18] This can explain the cooperative behaviour of those species which help to raise the young of close relatives.

I should stress that Hamilton was interested in solving an evolutionary enigma; he wasn't particularly concerned with human affairs. All the same, the human inclination to favour relatives is so glaringly obvious that it couldn't pass unnoticed. Everywhere in the world, those in high office who can get away with it do this on a grand scale. We call it nepotism, from the Latin for nephew—which is what past popes called their illegitimate sons.

Animals don't need to favour their relatives deliberately in order for kin selection to be at work in evolutionary time. If a way of life is such that individuals are often in the company of close relatives then, inevitably, close relatives will often be the interactors whether they realise it or not. In this case, one has to think statistically. On the other hand, there have been dozens of studies to show how animals (and other organisms) are able to recognise close relatives unconsciously. A variety of animals take their cues from the smell of their nest mates.[19] Recognition systems must have evolved very soon after life began, because at the molecular level our body cells are distinguishing between 'us and them' all the time. When the immune system makes mistakes it can accept dangerous outsiders or it can reject its own kind and cause autoimmune disorders, such as rheumatoid arthritis. Humans normally do recognise each other, of course, and we are well aware of it when we favour our relatives, but the evolutionary question is: Why should we want to do this? What has given nepotism its appeal? It is impossible to plot the course of its evolution, but conscious planning certainly was not a prerequisite.

DOING UNTO OTHERS

There is no problem in explaining the evolution of cooperation provided it improves the fitness of those who do it. It becomes hard to explain how selection can work only when it appears that a particular behaviour would reduce an animal's fitness. In cases where kin selection is unlikely, one possibility is that what appears to be helpful behaviour may not be altruistic at all. It might really be a reciprocal arrangement—a case of mutual help to nobody's detriment. As everybody knows, it is common practice all over the world for people to cooperate with each other for as long as both sides keep good faith. When cooperation becomes too one-sided it is not likely to last long because the arrangement, whether it be sharing meat or sending Christmas cards, will break down. Humans think about such things. The question here is whether and how this kind of behaviour could have evolved in non-humans.

It is a question that has fascinated games theorists for a couple of decades. In principle, although everyone would do better if everybody cooperated, the genes of a selfish individual would come out ahead in a population of cooperators that hadn't enough brain power to understand free-loaders. Cheats could therefore invade a society of cooperators so that cooperating would not be an *evolutionarily stable strategy* (ESS).[20] And yet there can be a point at which cheats become so numerous that the group as a whole becomes less efficient, so we must consider the balance of cheats and cooperators within groups. This brings us back to the difficulties of group selection, which, as I said earlier, is still a contentious area.

The problem of how cooperation evolved is commonly described in terms of a game of Prisoners' Dilemma. This comes from the idea of two prisoners being interrogated in separate rooms about a crime they did together. Let us assume that if both cooperate, and stick to their story, both will get a light sentence. If they blame each other with a version of the truth, both will be punished with a somewhat stiffer sentence. If one cheats by blaming everything on his faithful accomplice, the cheat will get off with a tap on the wrist and the other will get the worst punishment to be had. Under these circumstances, how could cooperation arise by natural selection?

Theory held that in single encounters it would pay to be selfish, but with repeated interactions, if each individual based its behaviour on the previous behaviour of its opponent, it would be possible for cooperation to evolve. This simple strategy became known as tit-for-tat (TFT), and it couldn't be beaten by the games theorists. That was a long time ago. It has since been shown that TFT would not be an ESS if there were genuine mistakes in judging an opponent's behaviour. Cooperation would break down. What about allowing for one mistake and withdrawing cooperation only after two defections? Would tit-for-two-tats (TF2T) be an ESS? Or what if . . .? Never mind. The whole thing has become a computer game with a life of its own.

In our daily lives, in one way or another, we are always playing Prisoners' Dilemma. Without necessarily thinking about it we monitor each other's behaviour and assess each other for reliability and worthiness of trust. And, privately, we make our decisions. Among cooperating wild animals, however, it has been extraordinarily difficult to identify and assess the costs and pay-offs that might be involved. For this reason there has been widespread interest in what appears to be a clear case of reciprocal altruism in vampire bats. These small mammals need a meal of blood almost every night. They cannot survive for more than two days without food, and they gorge themselves whenever they get the opportunity.

They like to feed on horses, usually hanging from the mane while they get the blood flowing with their sharp little teeth. It is said that they can tap into a human big toe without the sleeper feeling a thing. Back at the roost, the bat may regurgitate some of the blood for a companion who didn't get a meal. This really does appear to work on a tit-for-tat basis, although, even here, relatives and close contacts get preference over strangers.[21] Other clear examples have been hard to find because what looks like reciprocal altruism, a tit-for-tat arrangement, might just be cooperation for mutual benefit at the time; it may be that both cooperators are doing better by cooperating, even if one is doing a lot better than another. This presents no evolutionary problem; cooperation would be part of the optimal strategy for self-interest and would be favoured by natural selection.

As one analysis puts it: 'Reciprocity remains one of the most beguiling concepts in behavioural ecology'.[22] But so far there is no agreement as how important it is among non-humans.

None of this has much bearing on present-day wheeling and dealing, back-scratching or genuine acts of kindness and charity by modern humans. We have been considering only how cooperative behaviour might have evolved and we have to conclude that acts of genuine, one-sided cooperation between non-relatives are not easily explained, if indeed they do occur among non-humans. In any case, when most people think of altruism they think of conscious concern for others and a sense of what is good or bad, right or wrong and compassionate or heartless. They think, in a word, of *morals*. The question for sociobiology at this point is whether some form of animal cooperation can have paved the way for the uniquely human behaviour that we call altruistic in the moral sense. For a start, what does sociobiology have to say about the social mentality of humans and other species?

SOCIAL MODULES

Specialists disagree about the mind-reading abilities of chimpanzees. A chimp definitely has a concept of self, but can it understand that it knows something not known by another one? Can it understand that if it sees its keeper move a banana from one hiding-place to another, it will be privy to something that other chimps, who were out of the room, cannot possibly know?

Can it pretend to be hurt in order to deceive another chimp? The behaviour described in Chapter 8 suggests that chimps have some of this understanding—some theory of mind (ToM), as it is technically known—but it is difficult to measure, or even to demonstrate under strict experimental conditions. Let's give chimps the benefit of the doubt on the questions I've just asked. That would give chimps enough ToM, according to one study, to be compared with children of 3–4 years old.[23] Children who suffer from autism are likely to have a less well-developed ToM than others of comparable age. Autism has been described as a defect of empathy which makes it hard for the person to understand others' beliefs, desires and intentions.[24] If chimps had no ToM, then that kind of understanding would be unique to humans, having evolved in the human brain after the split with other apes. The prevailing view is that the capacity for a ToM in both chimps and humans may be located in a part of the brain called the lateral prefrontal cortex, which is present only in primates and is much bigger in humans than in chimps.[25]

Because ToM develops naturally in human infants at about the same stage in different cultures, it has been suggested that this is another example of a human mind module. There may even be an innate disposition to acquire a sense of fair play. That shouldn't be surprising in view of our universal tit-for-tat behaviour and the need for constant appraisals and reappraisals of our social relationships. People generally understand social situations more readily than abstract ideas—even when identical logic is involved. The well-known demonstration of this is based on a finding by Peter Wason at London University in the 1960s and followed up by others elsewhere.

To see how it works you could try showing your friends four cards with shapes drawn on them. One has a square at the top, one has a square at the bottom and the other two have circles, one at the top and the other at the bottom. You say that there are also shapes on the backs of the cards and it is a rule that any card with a square at the top must have a circle at the bottom. You then ask which cards must be turned over to see whether the rule holds good. Many people get it wrong. Most will get it right if you ask them to use the same logic in a social scenario. The rule this time might be that anybody holding a top-position pass must be a pensioner. If only one card has a pass at the top then only one card need be turned over to check the rule. All other cards are irrelevant—including any with 'pensioner' written on the front. In the first example, you need only turn the card with a square at the top. Tests like this, and more difficult ones, have led some psychologists to believe that we have a special ability for learning to monitor social fair play.[26]

Having a built-in cheat detector is still a long way short of having a social conscience and a system of ethics, but it is at least an evolutionary step in the right direction, and we haven't yet mentioned the emotions.

OUTWITTING THE BEAST: EMOTIONS AND MORALS

Fear, pleasure, surprise, anger, sadness and disgust are ancient emotions, and are shown in much the same way by people everywhere. In most cases, individuals from different cultures will correctly identify them from facial expressions in photographs.[27] Judging from behaviour and by comparing brain activity in laboratory studies, we share some primitive emotions with other animals too. It is as though we possess a few packaged responses that can be switched on as a reaction to any number of different situations. As with classical conditioning, we learn new triggers for old responses. Our environment, including the social environment, is constantly changing and is quite unpredictable, so evolution could do no more than prepare us with some degree of preprogramming based on common experiences over a vast period of time. In this sense a knowledge of the past, written in DNA, has been brought with us into the present. It is not knowledge that is accessible to the thinking parts of our brain, nor is it easy to control, and yet it helps to direct our present learning; it is much easier to become afraid of the dark, or a snake, or a spider, than it is to get scared of scissors or lawnmowers.

As one author puts it: 'Emotions are postcards from our genes telling us, in a direct and non-symbolic manner, about life and death'.[28] Stealthy footsteps in a dark alley, or a sudden bang; a mother's smile, or a writhing mess from the dog's inside; these will all arouse instant feelings, not difficult to imagine, which precede any thinking. Thoughts will come later, as will new and more subtle emotions. Some believe that the more subtle feelings are just a blend of the basics, as every shade is a blend of the primary colours. Others disagree. One way or another, when emotions are combined with conscious thoughts, new feelings emerge because it is impossible to think without any feeling.

And here is the root of our social emotions problem: feelings of loyalty, sympathy, and guilt are not like the primary emotions of fear or pleasure: they are more like feelings blended with learning and culture. They have different forms of expression as well as having different triggers. Unlike intelligence, our finer social feelings—our moral sensibilities—are not so easily explained by natural selection. We can accept that intelligence is adaptive because higher mental functioning would enable an individual to cope more successfully with social complexity, and with the unexpected in the environment. This doesn't strain evolutionary theory at all. But our moral sense is more difficult to account for, and it could be that it is not an adaptation at all. It could be that our distinctly human emotions emerged as a byproduct of other mental developments. In that case morality would be incidental, a spin-off from the distinctly human expansion of the brain. We have now strayed well into the realms of speculation and into an area where philosophers have more to say than biologists, but I have gone this far because I want to mention a shift in thinking that has emerged during the past decade or so.

Ever since Darwin there have been attempts to find a link between evolution and ethics, but the cold truth remains that natural selection is utterly indifferent to human values. As T.H. Huxley was quick to point out 100 years ago: 'evolution may teach us how our good and evil tendencies have come about; but, in itself, it doesn't tell us why we should regard this as good and that as evil, nor why we should strive for one and not the other'.[29] The philosopher David Hume had much earlier written that it is a naturalistic fallacy to think we can tell what *ought* to be from what *is*. And Richard Dawkins was only echoing this when he said he was 'not advocating a morality based on evolution'.[30] But a few philosophers are now looking at nature from a different perspective and have reopened the debate. One author, for example, believes that the naturalistic fallacy has outlived its usefulness and that modern evolutionary theory can support a case for a more egalitarian society.[31] Others are not convinced that evolution holds any ethical lessons but are having another look, if only to bring their arguments up to date.

Time will tell whether modern evolutionary ethicists can make much headway. They have a fundamental problem, in that our brains are much the same as they were thousands of years ago but our societies have changed radically and can change again within a generation. While it is comparatively easy to link ourselves, as individuals, to our evolutionary past, it is far more difficult to relate the distant past to modern societies. Evolutionary thinking was able to slip comfortably into the medical profession, because any doctor's waiting room is likely to be a parade of evolutionary consequences in the form of lower back problems, impacted wisdom teeth, and circulation deficiencies ranging from haemorrhoids to pregnancy leg.[32] It may not be much comfort to the patients to know that their ancestors had more accommodating jaws and once had sturdy, straight backs supported by four limbs, but at least it provides the profession with a sound theoretical framework through which to view the human body. Evolutionary psychology is the equivalent framework for the mind-doctor but for a variety of reasons it is a more controversial one and there may never be a finalised version of it.

Even so, the application of evolutionary thinking to human social behaviour, whether we call it evolutionary psychology, sociobiology or something else, is here to stay. Specialised journals on the subject were established long ago, textbooks are now available, and a growing number of university departments not only support research but provide courses for undergraduates. It is a field of study destined always to be plagued by controversy which can never be settled by fossil evidence. There will always be some enthusiasts who are prone to overstate their case and others who like being controversial. As for the social sensitivities involved, they can be guaranteed to generate strong emotions in all kinds of people for all kinds of reasons.

Among those who denounce the whole business are those who are still unable to accept that our minds have been moulded by the common experiences of our distant ancestors. Surely, they argue, we are at least aware of our emotions even if we cannot always control them. We are at least privately aware of the forces that are pushing and pulling us as we try to make up our minds, aren't we? Well, possibly not, as I hope to make clear in the next chapter.

CHAPTER 15

OLD BRAINS, NEW BRAINS

F OR EARLY ANATOMISTS, AN opened skull must have made about as much sense as a bowl of yogurt. It took nineteenth-century microscopes to reveal beyond doubt that the brain had a cell structure of any kind. We now know that each nerve cell, or neuron, has a cell body with fibrous extensions (dendrites) coming from it that look like the roots of a plant. There is typically a single long fibre called an axon which carries the nerve impulse away from the cell body to neighbouring cells. By biological standards a nerve impulse travels quickly, but electricians will not be impressed: a current in copper wire moves about a million times faster. Axons are usually insulated with a white, fatty sheath so that tissue which is mainly fibrous looks white against the greyer appearance of tissue in which cell bodies predominate. Nerves are actually bundles of axons, so they look white too. The axons which carry impulses away from the central nervous system to distant muscles can be metres long in some animals, but brain cells are so small that one human brain contains over 100 billion of them—close to 20 cells for every person on Earth. On average, a growing human brain must generate about a quarter of a million neurons a minute in the nine months before birth.

On this microscopic scale, brain cells show an enormous variety of shapes, sizes and molecular differences. Some are so densely branched that they look like leafless shrubs. Others, which communicate with different parts of the brain, have longer fibres with fewer branches. To some extent brain cells change with experience, for it is the cells, rather than brains as a whole, which learn, remember and forget. In general, brain neurons don't regenerate, so they have to live as long as we do. We also lose a great many cells early in life because we grow more than can possibly survive. Selective cell death is part of normal brain development.[1]

In passing from one cell to another, the nerve impulse (action potential) has to jump a gap between the axon terminals of one cell and spines on the dendrites of the next. The gap, or cleft, occurs at specialised points of contact called synapses, and there could be as many as 100 trillion of them in a human brain because each neuron could be in contact with many thousands of synapses at any one time.[2]

Nerve impulses jump synaptic clefts chemically, not electrically. When an impulse arrives at a synapse it causes molecules of a transmitter substance to be released. These neurotransmitters bind to specialised sites on the receiving terminals and alter the membrane's ion permeability one way or the other—either exciting it or bringing it back towards its resting state. It is not an all-or-nothing effect; the changes are graded. And as there are thousands of synapses on the dendrites of a neuron, some inhibiting and others exciting, any new action potential will depend on whether all the added and subtracted inputs produce a big enough stimulus. If so, the receiving cell will trigger a firing at the base of its axon. After neurotransmission, chemicals known as second messengers can modify receiving cells permanently, possibly as part of a learning process.[3]

Neuroscience has come a long way since Galen and his four bodily humours, but we have come most of the way in a short time. Not until 1921 was it shown that any chemicals were involved in transmitting nerve impulses. There are now known to be nine classical neurotransmitters and perhaps 50 more subsidiary ones, which have different actions depending on which of the main ones they accompany.[4] Some of them act as hormones in other parts of the body.

The brain actually produces more chemicals than any other organ and is thought to express some 30 per cent of our genes. All this delicate chemistry is safeguarded by a fatty barrier in the brain's finer blood vessels which prevents interfering chemicals from getting through. But, being what we are, we humans have discovered a long list of mind-altering chemicals—from alcohol to Valium—which, for better or worse, can cross the barrier.

Basically, nerve impulses are the same in all animals. Much of the pioneering research was done on squids, which have axons a hundred times thicker than any human ones, making them easier to dissect out and experiment with. It is less easy to generalise about whole brains. Some invertebrates have little knots of nerve tissue (ganglia) arranged throughout their central nervous systems without any one being in overall control. Others, such as earthworms, have a dominant ganglion at the head end but it is not indispensable. A worm with its head chopped off will be kept alive by other ganglia until new forward segments regenerate. Vertebrates also have ganglia in the nervous system (the solar plexus is a cluster of them) but every vertebrate has an obvious brain with specialised nerve tissues and, although the loss of some chunks can be survived, the different parts are never entirely interchangeable, and natural regeneration of damaged brain tissue is very limited.

LAYOUT OF THE BRAIN

In all vertebrates the embryonic brain develops from a flat sheet of tissue, called the neural plate, which extends as a strip down what will become the animal's back. The strip sinks deeper into the embryo and the edges roll up along its length to form a neural tube. In humans this happens at the end of the third week after conception when the embryo is about three millimetres long. The hollow tube swells at the front to form the brain, while the rest of the tube becomes the spinal cord. The skull and backbones are laid around it, first as cartilage, and later in the calcified, jointed pattern of the bones. After four weeks the embryo can be recognised as mammalian but the neural tube is still open at both ends and the brain consists only of three bulges, which represent the three regions of all vertebrate brains: the forebrain, midbrain and hindbrain.

In the cold-blooded vertebrates the brain remains in the shape of a closed, lumpy tube, so the three sections, with their further elaborations, remain more or less one behind the other and easy to identify. The forebrain bulges out on both sides as the cerebral hemispheres, which deal with the sense of smell among other things. The midbrain is largely devoted to vision (optic lobes) and the roof of the hindbrain swells up as the cerebellum, which is mainly responsible for balance and coordinating the impulses which pass up and down the spinal cord. The original cavity of

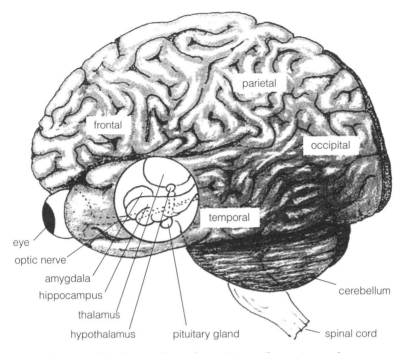

In this drawing of the human brain the positions of some internal structures are shown within the circular window

the neural tube persists as a series of spaces (ventricles) filled with cerebrospinal fluid.

I should mention at this point that the walls of the forebrain, at the rear end, become thickened into specialised structures. One of them, the thalamus, is a major relay centre. Below it, the hypothalamus is involved in just about all behaviour common to mammals, including the expression of emotions. It communicates through a stalk with the pituitary gland, which is really a double gland, only half of which comes from brain tissue. In humans it secretes nine major hormones, and we couldn't survive without it.

The term 'brainstem' gets heavy use in the literature. It refers to those parts formed from the tubular walls of the hind- and midbrains, but authorities differ about including the cerebellum.

The same basic plan is common to all vertebrates despite a lot of variation in sizes, shapes, and relative importance of the different parts. All classes except mammals have smooth cerebral hemispheres. A section through those of a bird shows white stripes where tracts of fibres from the thalamus and brainstem pass through the grey matter. It is a wiring system characteristic of creatures whose behavioural programs are relatively fixed.

Mammal brains took a different direction. In mammals a new blanket of grey matter, called the neocortex (Greek *neos*, 'new'; Latin *cortex*, 'bark'), forms over the surface of the cerebral hemispheres. It comes from cells which, in ancestral reptiles, would probably have been located in the same relative position, though cells from other locations could have contributed.[5] The origins of brain parts are traced by studying the development of embryos, by comparing tissues of different living species, and by comparing the genes expressed in the different brain tissues.[6]

The human neocortex is not much thicker than a matchstick but it amounts to a large proportion of the brain. The bulk is made possible by having the brain surface thrown into deep grooves and ridges. A few mammals (including rats and rabbits) have smooth cerebral hemispheres, but most have a highly convoluted surface which, on a human foetus, begins to show up after the sixth month. All primates have the area of the neocortex further extended by the overgrowth of the cerebral hemispheres backwards and sideways over the rest of the brain. In humans the overgrowth is so extensive that, from above, the entire brain looks like a soft, grey walnut.

All human brain surfaces have the same main features, but the detailed pattern of grooves and ridges is highly individual. I was about to say that brains are like faces in this respect, but it wouldn't be strictly true. The two sides of a face normally look more alike than the two halves of a brain. On the surface, human cerebral hemispheres are not even a matching pair. The evolutionarily older structures can only be seen from underneath, or by cutting.

MAINLY HUMAN

After birth the human brain grows in spurts, during which some neurons are lost and others grow bigger and multiply their numbers of dendrites, spines and synapses. The last growth spurt occurs at 14–16 years. Normal development depends on stimulation of the neurons; deprived babies and experimental animals reared in solitude with nothing to exercise their senses don't develop normally. Newborn kittens which had one eye experimentally covered failed to make proper connections among the neurons in the visual pathway from that eye. The result was that, during a critical period, axons in the path of the uncovered eye grew to occupy extra space in the visual cortex.[7] Targets for neuron growth are set by genes but, as with behaviour, DNA provides only the framework; the fine-tuning remains flexible, and brain development is a matter of using it or losing it.

The thought of a scalpel in the living brain might provoke a shudder, but the brain has no pain receptors and investigations can be done on fully conscious patients with their brains exposed under local anaesthetic. In the past, patients who had suffered brain damage, especially from the ravages of World War I, were the main source of new information on how the human brain is organised. Today's specialists can learn a lot about brain activity by using scanning devices based on radioactive emissions or magnetic fields. One technique, called TMS (transcranial magnetic stimulation), can disrupt neuron firing in small regions of the brain surface down to about two centimetres deep so that the consequences of interference can be studied without so much as a scratch.[8]

No part of the human brain is unique. Thirty years ago it was popularly described as a triune organ in which the modern version had been added to an earlier mammalian structure, which in turn was founded on the primitive reptilian plan. The idea of a three-in-one system is not entirely unrealistic but it would be wrong to think that earlier brains are embodied, unchanged, beneath our neocortex, or that brains evolved in an additive, stepwise fashion. All vertebrate brains are modifications and developments of the original model. The major part of our hindbrain is still the cerebellum and it is still mainly concerned with balance and coordination, though it also plays a part in learning of the classical conditioning sort. In contrast, the midbrain, which in fish can be the most prominent of the three regions, is tiny in people. The roof of it is represented by two pairs of little swellings. The forward pair still receives some input from the eyes but the information is relayed via the thalamus to the neocortex. Our thalamus remains the main relay centre for all sensory information (except the sense of smell) reaching the neocortex.

The simplest map of the neocortex would divide it into four regions (lobes) which take the names of the skull bones that overlie them, but these broad divisions into frontals, parietals, temporals and occipitals don't correspond to neat divisions of cortical labour.

Deeper within the forebrain there has been some major reorganisation. Structures that formed very early in brain evolution have been overgrown by newer tissue so that they now lie in a band above the brainstem—sandwiched, in a sense, between the old and the new. Collectively, these structures are now thought to be a functional unit, which has come to be known as the limbic system (Latin *limbus*, 'border'). The hypothalamus is a central part of it and perhaps the most indispensable component, although it doesn't amount to much in size. Students can expect to be told that the limbic system is the visceral and emotional brain, responsible, among other things, for the four f's of life: feeding, fighting, fleeing and . . . copulating. More recently, it has been found that the limbic system also appears to link emotion, particularly fear, to long-term memory.

LEARNING AND MEMORY

Fear conditioning occurs in all mammals that have been studied, and possibly it occurs in all vertebrates. It seems that some ancient pathways have been conserved in the brain over a very long time. There is obviously adaptive value in having a rapid-response circuit for perceived danger signals. First we jump, and then, perhaps with a little embarrassment as the signal reaches the visual cortex, we realise that the snake in the woodpile is just a piece of old pipe, or the spider in the bath a plastic joke. Emotional memory is believed to be located and stored in durable circuits in the limbic system. It is the result of emotional learning based on fear conditioning. It very likely operates independently of our conscious awareness and some emotional memories may be totally inaccessible to our conscious memory.[9] The question how far human behaviour is ruled by emotion can itself raise some emotions. There is, of course, no simple answer, but every journalist and politician knows that no matter how the facts and figures of an issue may stack up it is the emotive bit that grabs public attention.

The idea of subconscious motivation was easily accepted by the Freudians, but Sigmund Freud (1856–1939) had absolutely no experimental evidence for his ideas about how the subconscious is organised. Theories of infant sexuality, conflicts of id and superego and so forth are taken with a pinch of salt by hard-nosed, statistically minded psychologists, but nobody is seriously challenging the evidence that not all parts of the brain are accessible to our conscious, reasoning minds. The relationship between conscious and unconscious, and between the older and more recent parts of the brain, is only dimly understood and it is fair to say that there are mysterious lines of communication.

Frontal lobes

Until after the middle of the twentieth century few researchers believed that memory could be located in any particular part of the brain, though it was long believed that the frontal lobes, which extend well back from the forehead, were the seat of higher intellect, even though accidental damage to this area caused a lot more than intellectual loss. The most publicised accident story is that of Phineus Gage, a railway worker who, in 1848, happened to strike a spark while ramming powder for rock-blasting, and blew an iron tamping-rod clean through his head. The heavy bar entered his left cheek and shot out from well behind his forehead. Before this happened Gage had been of ordinary intelligence and was considered to be a well-mannered and considerate person. Afterwards he was said to be capricious, inconsiderate, irreverent and given to uttering gross profanities. Gage still had his memory and was still intellectually capable of doing his job, but he had undergone a profound change in personality.

There have since been many case studies, as well as years of experimental research, to assist in mapping out the functions of different parts of the neocortex. The frontal lobes, which amount to 20 per cent of the total, have several regions which are connected functionally to other parts of the brain. Damage to different areas of the frontal lobes can cause symptoms ranging from loss of control of fine motor movements to altered sexual behaviour. Social and sexual behaviour, calling as it does for flexible responses to cues and feedback from other individuals, must make heavy demands not only on our sensitivities but also on learned rules about what is acceptable behaviour. In turn, that makes demands on memory. Particular parts of the frontal lobes are believed to be responsible for selecting behaviour that is appropriate to the immediate situation in the light of existing knowledge and with regard to memory stored in the temporal lobes.[10] Extensively damaged frontal lobes can lose this capacity, so that affected patients lose much of what it takes to be socially acceptable. They may show no concern about such behaviour as sitting in urine-soaked clothes, making personal remarks to strangers or masturbating in public.

Research with monkeys indicates that they can suffer equally dramatic changes in social behaviour as a result of frontal lobe removal. After the operation, dominant males that were returned to their colonies were able to hold their high rank for only a couple of days, after which time they were deposed and became social outcasts.

Temporal and other lobes

The temporal lobes are at the sides of the head. They have long been associated with memory, though there are specialised areas there for analysing information from the eyes and ears as well—including language and music. Memory is localised in the temporal neocortex and in the hippocampus and amygdala which lie below it as parts of the limbic system. In 1953, a patient now famous in the textbooks as H.M. underwent surgery to remove these parts on both sides in an attempt to control a severe and worsening case of epilepsy. Afterwards, although he could recall his childhood and other distant events very well, H.M. had scarcely any memory for events following the surgery. Despite retaining above-average intelligence and remaining cooperative and well-mannered, he could master only simple, routine jobs because everything he did was like a new experience for him. After six months of monotonous employment he still couldn't describe to others what it was he did, what the workplace was like, or by what route he was taken there every day. Yet his vocabulary and conversational ability were still intact. For H.M., every day was like a new awakening in an unfamiliar situation.

The case of H.M. led to hundreds of experiments on laboratory animals, and the current interpretation is that there are different classes of memory with different storage places. We have seen that emotional memory is different from conscious recall, and it is also well known that movement patterns (motor learning) are different again. H.M. could scarcely learn anything that required recall because he couldn't build on what had gone before. But he was able to improve motor skills with practice even though he couldn't remember ever having practised them.[11] This separation is consistent with the folk wisdom that memories of our physical skills, like riding a bicycle or swimming, take care of themselves. There is evidence from many sources that the cerebellum is involved in that sort of learning.

We are all familiar, too, with the idea that we have a short-term or working memory that is separate from that which we hope to keep on tap for weeks or years to come. Contrary to what was once thought, the one is not converted to the other over time: the two systems are independent and can be processed simultaneously. Probably there are a number of kinds of short-term memory. Short-term deficits can result from damage to the temporal, parietal (top of head) or frontal lobes. Some patients can remember lists of words if they hear them but have trouble with the written form. For others the opposite is the case. Still others can recall both kinds of stimulus but they get the order muddled. Long-term memory certainly exists in different forms and is stored in different areas. Parts of the hippocampus, as we saw with Clark's nutcracker, are responsible for spatial memory, which accounts for H.M.'s inability to remember travel routes. This observation also fits nicely with the recent discovery that London taxi drivers, who can only become licensed by passing a notorious test on city traffic routes (a feat known in the trade as getting 'the knowledge'), have an enlarged part of the hippocampus which swells with experience.[12]

Exactly how and why some brain regions show physical change with use is still being intensively studied, but several kinds of change have been shown since the Canadian psychologist Donald Hebb proposed that learning must modify the synapses.[13] When animals are trained to do a particular task, some neurons grow more dendrites and acquire bigger spines, more synapses and more of the vesicles that hold transmitter substances. As the absolute size of a brain is fixed, any expansions must presumably be accompanied by reductions elsewhere. Even so, the old adage 'use it or lose it', which takes effect early in life, evidently continues to hold good until death or senility put an end to the learning process.

SEEING IS BELIEVING

The occipital lobes, also known as the visual cortex, occupy the back of the head, and as far as is known they deal exclusively with information coming from the eyes.[14] Starting from the idea that the eye was like a camera, it was once thought that light images from the retina were transmitted to higher levels of the brain where another process made sense of them. This idea of separate processes for sensing and understanding persisted until the 1970s, after which the idea became untenable. The current view is that no part of the specialised visual area serves only to relay information, and no single master area of understanding can be found. Instead, all visual parts of the brain (the retina can be regarded as part of the brain) are actively involved in transforming the incoming signals and making some contribution to perception.[15]

There is a lot of cross-referencing and integration but, generally, the further along the visual system a cell is located, the more complex a visual stimulus must be to excite it. For example, cells in the retina are excited by spots of lights; but after relay through the thalamus the signals are projected to different parts of the visual cortex where cells will respond only to colour, or to bars of light at certain angles, or to movement in certain directions. PET (positron emission tomography) scans of human brains show activity in different regions according to whether a person is looking at a colour pattern or a black-and-white pattern, and whether there is also movement. Less specialised cells in other areas respond to combinations of these stimuli.

There are brain-damaged people who are not only condemned to seeing the world in shades of grey but who cannot imagine or understand colour, even though they once had normal vision. Perhaps the mental image shares the same part of the brain as perceptions of the real thing. Other victims of brain damage cannot see or understand movement. For them, moving objects simply vanish from one place and reappear where they stop. Flowing streams of liquid look frozen.

The symptoms of one patient, who suffered stroke damage to part of the non-primary visual cortex, provide a vivid demonstration of limited perception. He could see lines and angles and simple shapes but he couldn't make sense of them. He could laboriously copy a drawing of a cathedral without being able to see or understand that what he had drawn was a building.[16]

One intriguing evolutionary possibility is that our brains have become specialised for recognising faces. In the opinion of some experts it might just be that we have a specialised area for processing complex patterns, be these facial, geometrical or whatever. I have no counterargument, but I feel sure I could learn to recognise faces a lot faster than I could learn abstract patterns.

LEFT BRAINS AND RIGHT BRAINS

The cerebral hemispheres are not entirely separate. They are joined by a few hundred million fibres in a thick band called the corpus callosum (Latin, 'tough body'). There is a much smaller cross-connection further forward and another link which joins the hippocampus with its partner on the other side. Through these, and some minor bridges, most parts of the two hemispheres are connected. Given such ample provision for the two halves to communicate, one might wonder why we hear so much about the influence of one side or the other on personality and aptitude. Do we really have a logical, analytical left brain and a more creative and intuitive right one? In a word, no. Well, not as separate halves. There are differences, but the two sides work to complement each other. Details of the relationship are likely to vary with the individual, the sexes, and the type of information being processed.

There are plenty of theoretical ideas precisely because the paucity of hard evidence allows for them. Even the relatively well-founded conclusion that language is located on the left is by no means cut and dried. In nearly all right-handed people, and some two-thirds of left-handers, the main elements of language, such as grammar and vocabulary, are processed on the left side but other aspects, such as intonation and emphasis, will light up the right side in scans.[17] On the strength of the data, Professors Kolb and Whishaw cautiously conclude that in right-handed people the left hemisphere has the greater role in language, writing, arithmetic and verbal memory, while the right hemisphere mediates spatial functions, pattern recognition, faces, sense of direction, geometry and non-verbal memory, including music.[18] There was a time when handedness was thought to be as uniquely human as language, but that idea has been quashed by the evidence. While it is true that humans are more predominantly right-handed than other species, many animals, even toads, show an individual preference for one hand or the other and have asymmetrical brains to some extent.[19] Quite possibly, brain asymmetry has progressed in humans because selecting for more tissue would have selected for energy saving as well. There is less tissue involved in having one specialised area with cross-wiring than there is in having duplicate areas.[20]

If the two halves of the brain were not able to share information it would obviously be much easier to see which side of the brain handles what. This condition arises naturally, but fortunately rarely, when a child is born without a corpus callosum. It can also be created by surgery when drastic measures are called for in dealing with epilepsy. Operations that severed the corpus callosum were first carried out in the 1940s, and there have now been many studies of split-brain patients.

If a split-brain patient is allowed to see an object through only one eye then only one brain hemisphere will get that information. Each side can interpret information from printed words, but both halves are needed to put separated meanings together. A patient who had the word 'sky' flashed to one hemisphere and the word 'scraper' flashed to the other could only respond with a drawing of clouds and a scraping tool, not a tall building. In most people the isolated right hemisphere has scarcely any capacity for spoken language and is severely deficient in problem-solving. In contrast, the isolated left half can cope perfectly well in this regard. Split-brain patients who are shown an object only to the right hemisphere have no way of talking about it or even naming it— though they know full well what they are looking at. There are always exceptions: one patient who had left-brain damage resulting from a stroke found the capacity to talk from the right brain after 13 years.[21]

I think the most fascinating revelation of split-brain studies is the phenomenon of false memory. We are all prone to make errors when we recount a story from memory, as all police officers must know if they have much experience of interviewing eye-witnesses. It is as though we construct a model or schema for an incident and then our memories become selective so that we remember details that fit the schema. Some of these memories may be false. With split-brain patients it is the left brain that generates the false memories; the right brain is more truthful. I should emphasise that the patients themselves have no suspicion that their memories are deceiving them.

In experiments done with split-brain patients during the early 1980s, each hemisphere was tested separately and simultaneously by presenting two sets of choices, one to each eye. The patients were asked to select, by hand, one of several small pictures to match up with a larger picture above. Each hemisphere made sensible matches. In one instance the left brain matched a chicken's foot with a chicken's head, while the right brain matched a shovel with a snow scene. The patient was then asked why a shovel had been selected. Now the two hemispheres could not share information and the left brain had not been able to see the snow scene. But only the left side could talk, the right brain having no capacity for language. The left brain's reply came without hesitation: 'to clean out a chicken shed'. In similar experiments, subjects' right brains were shown pictures of nudes. One lady blushed and began to giggle, covering her mouth with her hand. Asked what was the matter, the patient felt pressed for a rational explanation. Her left brain told the researcher that he must be using some sort of machine that affected her.[22] In practice these experiments are complicated by the fact that information from the eyes crosses over in the corpus callosum so that the right hemisphere processes data from the left visual field and vice versa. The same applies to hand and finger movements, so that the right hand responds to the left hemisphere and the left hand responds to the right hemisphere. We needn't dwell on this, but it obviously determines the way the experiments are set up.

Split-brain patients are very unusual people but they do demonstrate what the brain is capable of in terms of self-deception, and they give some substance to the claim that memories can play tricks on us.

MALE AND FEMALE

There are some differences in mental functioning between the average male and the average female, although the variations between individuals, irrespective of sex, are vastly greater. Girls from about the age of eleven show better language ability on such tests as filling blanks to make words and sentences. Girls are also better at some perceptual tasks, such as spotting matching patterns or seeing that something has been shifted or removed from a pattern. Boys do better at those visual puzzles that involve spatial judgement, such as map-reading and targeting skills, and mentally rotating objects to visualise them from different angles. Boys are also better at some mathematical problems, though girls are better at straightforward calculation. Finally, males of many species are more aggressive. In boys this difference begins to show up at two or three years and persists into adulthood. Various explanations are offered by different writers, none of them being mutually exclusive. All the old nature–nurture arguments get an airing in the literature, but most specialists would probably concur with Kolb and Whishaw, who find 'no evidence that the observed sex differences in verbal and spatial behaviours can be accounted for solely on the basis of environmental or social factors'.

There is no question that hormones play a major role in the development and functioning of the nervous system. The recipes for male and female development, laid down on the X and Y chromosomes, include instructions for the timely production of male and female hormones, and these in turn influence the sexual characteristics of the brain. It isn't completely understood how hormones have their organising effects in the nervous system but whereas the effects during development are usually permanent, those in adult life are typically related to reproductive cycles.

When all goes well, a human embryo during the second month will grow male or female gonads (testes or ovaries) internally. On the outside the embryo will develop a genital tubercle and a pair of folds flanked by swellings on either side. At three months (when the embryo is about 30 mm long and is more correctly called a foetus), these components begin to take shape one way or the other. If a Y (male) chromosome is present, the gonads will be testes and they will begin to release male hormones. If this doesn't happen, or if for some reason the hormones cannot act on the target tissues, the default form will be female, despite the presence of the Y chromosome. The tubercle remains as the clitoris, the folds become the inner lips, and the swellings become the outer lips of the female genitalia, as if no male chromosome had been present. In the normal course of events for males, the tubercle and folds combine to form a penis and the swellings fuse to form a scrotum.

Hormones, as messenger substances, are involved in some complicated chemical relays which start at the hypothalamus and pituitary gland. In order for the male genitals to develop properly, the hormone testosterone must be converted into another substance by an enzyme. In one remarkable genetic condition, a male foetus is deficient in this crucial enzyme and development of the genitals is delayed until the tissues can respond to testosterone itself. This happens at about twelve years old. The condition is known (from the name of the missing enzyme) as 5-alpha-reductase deficiency and it was first recorded among related villagers in Columbia. The Spanish-speaking locals call affected children *guevodoces* ('penis at twelve'). Although they look like girls, these children are distinguishable and are reared in the expectation of a spontaneous sex change at puberty.[23]

Not only do sex hormones act on the genitals, they also affect brain development so that mind and body are harmonised. When the programming fails, a person can grow up to feel that he or she has the body of the wrong sex. This condition, called gender dysphoria or transsexualism, is distinct from homosexuality and is much rarer. Records from a specialist clinic for transsexuals in Holland show about 150 new cases a year over a 25-year period. Of this intriguingly constant number, about 90 each year were treated with hormones and surgery.[24] Not surprisingly, there appear to be critical or sensitive periods in brain development, at least in laboratory animals. At the sensitive time the hormonal effects 'appear to extend to all known behaviours in which males and females differ'.[25] In humans, a few naturally occurring conditions provide supporting evidence. There is a genetic defect which causes baby girls, around the time of birth, to be exposed to masculinising hormones.[26] These girls grow up to be more tomboyish and aggressive than their unaffected sisters. It isn't just a case of different upbringing; the effect shows up even in the infants' choice of toys. In any case, mothers of affected children would be unlikely to try to make their affected girls more masculine.

Exactly how hormones influence behaviour is still to be explained, but one possibility is that male and female brains are made more or less responsive in different environments. It was discovered in the mid-1980s that exposure to gonadal hormones around the time of birth could determine how stimulation from the environment would alter the growth of dendrites in later life. It is also known that the female hippocampus is much more plastic in this regard than the male's.[27] In female laboratory rats the dendrites in parts of the hippocampus actually grow and lose spines quite markedly in time with the rat's four-day cycle, and, as Kolb and Wishaw point out, there is no reason to doubt that similar changes occur in time with the human reproductive cycle. There are evidently brain changes of some sort at work; it surely cannot be coincidence that nearly half the crimes committed by women are carried out during the four days preceding menstruation.[28]

Questioning the significance of sex-related brain differences brings us back to speculations about our ancestral lifestyle. I can only comment that the spatial and targeting tasks of the average man, and the language and perceptual skills of the average woman, do fit very well with plausible divisions of labour. If men were roving far from base in the course of hunting (or scavenging kills), while women and their children were foraging and gathering what was edible from among the vegetation, then the mental differences that can still be detected in modern humans are not really surprising.

THE ULTIMATE ENIGMA

For dualists, like Descartes, consciousness, or mind or soul or whatever we choose to call it, was something separate from the brain. Modern scientists with reductionist leanings are content to regard consciousness as an emergent property of brain activity— something that happens when enough neurons fire in synchrony. A development of this idea, admirably explained in clear language by Susan Greenfield, a professor of pharmacology at Oxford, is that there is a continuum of consciousness, reflected in the sophistication of the brain. In this view, a monkey is more conscious than a snail and an adult person is more conscious than an infant, who is more conscious than a foetus. The level of consciousness is determined by the number of neurons that can be recruited, or corralled, to fire together.

The brain of a newborn baby has very modest connections with relatively small assemblies of neurons, which can be synchronised. If a toy is hidden from view it ceases to exist for the child, as it does for many animals even as adults. Out of sight is out of mind. More mature humans can continue thinking about a hidden object and have memories and experience to call on.[29]

Other workers have tried to be more specific about the state that synchronised neurons are in when they are giving rise to consciousness, but philosopher critics protest that details of that sort can only help us to know what *produces* consciousness; the hard question is what *is* it? In reply, the neuroscientists can quote William James, who said over a century ago that consciousness is not a thing, it's a process. If any philosophers are satisfied with that, they might retort that if neural processes give rise to this wonderful cinema in the skull that we call consciousness, then who or what is watching the film? The observer cannot also be the observed. This last point is not the conundrum that it used to be because we now know that the brain isn't just a single lump: it has different structures and structural levels.

Wolf Singer is Director of the Max Planck Institute for Brain Research in Frankfurt. He suggests that the same processes which enable an animal to deal with information from its senses are used to deal with the results of that processing. That is to say, there are second-order processes which treat the output of the first-order processes in the same way as these deal with signals from the senses. Human brains can therefore be aware of being aware, and know that others are aware of our being aware and so on. In this view, complex brains are capable of perceiving their own functions, not because they have anything special but because they have more of the same equipment devoted to higher-order functions.[30]

For Singer, and many others, the process *is* the sensation, but he doesn't believe that our higher consciousness could emerge from an isolated brain in the glass tank of science fiction. One has to be conscious *of* something and individuals possess a vast circuitry of knowledge, experience and memories from social learning. This could only have been built up by communication between different brains.

Professor Igor Aleksander specialises in electrical engineering and neural systems at Imperial College, London. He argues that it is the sum total of memory and knowledge, and the continuity between them, which make up what we call consciousness. But he is quite optimistic that an artificial neural net, made up of silicon chips, could learn and remember well enough to meet the consciousness criteria.[31] The thought of this will be stimulating to some but will make others groan.

Groaners are not likely to be comforted by some of the alternative ideas. Roger Penrose, a mathematics professor at Oxford, argues that the aspect of consciousness that we call *understanding* can never be handled by computers and that any explanation will involve venturing into the bizarre world of quantum physics, where waves and particles are said to be one and the same thing and can be in two places at once. Penrose suggests that the processing, if that's what it is, could take place in protein tubules that are plentiful *inside* neurons.[32] This won't do for David Chalmers, a philosopher with a mathematics background, who thinks that consciousness is so utterly different from anything that we are familiar with that it needs to be given a category of its own, like gravity.[33] At least one other philosopher, Mary Midgley, finds that idea singularly unhelpful.[34]

Perhaps, like dogs or chickens, we have limits to our understanding, and the puzzle of consciousness is beyond us. But I shall give the last word to Professor Antonio Damasio, head of the Department of Neurology in the University of Iowa. He is an optimist, and in his opinion 'it is probably safe to say that by 2050 sufficient knowledge of biological phenomena will have wiped out the traditional dualist separations of body/brain, body/mind and brain/mind'. I hope Damasio is right, for as he also remarks: 'the mind will survive explanation, just as a rose's perfume, its molecular structure deduced, will still smell as sweet'.[35]

CHAPTER 16

THE FUTURE IS COMING

T HERE IS NO REASON to think that our brains today are any different from the brains of people who lived 50 000 years ago. With our technology the people of the Upper Palaeolithic could have lived like us. Social evolution is Lamarckian in that it depends on the transmission of behaviour that has been acquired by learning, not by a change in gene frequencies. The arts seem to be no exception: the smile of the Mona Lisa would surely have been a lot less subtle had the pigments been applied to a cave wall with a stick, and Bach could not have written the Brandenburgs when there was no written music and the only instruments were drums and bone flutes. But the most conspicuous social change has come about in response to technological advances in such basics as food production, transportation, weaponry, buildings, medicine, the internal combustion engine and electricity. Tribal groups who were still in the Stone Age during the twentieth century were able to adapt to the electronic age in two generations. All very obvious, but I think we tend to forget how much we are creatures of our technology. Of course there are interactions and no doubt there have been points of critical social mass, just as there have been social revolutions and Dark Ages to influence learning and technological innovation.

There is an interaction, too, between social and physical evolution, because our technology has relaxed those selective pressures which otherwise would still be as important to us as they are to other animals. Compare the people of any modern society with the individuals of any wild mammal population and you will see that our way of life supports a far greater variety of physiques and individual differences in health, vitality and just about everything else you might care to measure. By comparison, the animals that breed in the wild have to pass through a very fine survival filter, which keeps detrimental variations within strict limits. If we still had to live as wild creatures, a great many of us, myself included, would never have survived to reach adolescence.

One of my problems was a troublesome appendix, which, had it not been taken out, could easily have burst and given me a fatal dose of peritonitis. The little cul-de-sac of intestine technically called the vermiform (Latin, 'worm-shaped') appendix is the vestigial end of a much bigger bag called the caecum. A functioning caecum is a fermentation chamber filled with bacteria which break down the cellulose in vegetation. Howler monkeys feed on leaves and have a big caecum. So do rabbits. Our appendix is a little stub of a thing, useless in digestion. Dogs have none at all. We would surely be better off without one (even if it does produce a little lymph), because an appendix blocked with partly digested food can be lethal. In a world without surgeons, mutants born without an appendix might be sufficiently favoured by natural selection for the appendix to be eliminated altogether.

One sometimes reads that we must have brought our physical evolution to a halt because we have suspended natural selection to the extent that we have. But nobody can be completely sure of that, because it has to be based on assumptions about the future. Organic evolution and social change work on different time scales. Nobody really knows what the human body will be like after another million years have passed. In predicting the technological and social future, even over a few decades, some of the best-informed people of recent generations have been laughably wrong. In view of all this, you may well wonder why I am bothering with this chapter at all. My answer, I suppose, is that I feel the need to finish up by drawing a few loose threads together, also to see which way we would be heading if present trends continued for long enough. Evolutionary change, after all, is based on the shifting frequencies of genes, so with the immediate future in mind it could be interesting to look at the gene-shifting forces that are operating at present.

POPULATION CHANGES IN THE MODERN WORLD

United Nations population projections show that in the next 50 years the world population will grow from the present six billion to somewhere between 7.3 and 10.7 billion, with 8.9 billion considered most likely. At present, with an annual growth rate of 1.33 per cent, we are adding 78 million people to the total every year, but this increase has a very patchy distribution.[1] For example, in the 1950s there were two and a half times more people in Europe than there were in Africa. Now there are more Africans than Europeans. In half a century, Africans have trebled their numbers, Asians and South Americans have more than doubled in number, but the North American population has increased by only 50 per cent and that of Europe by only 20 per cent. In the next half-century the European slice of the global pie chart is projected to shrink to 7 per cent (from 20 per cent in 1960), while the African portion is expected to do almost the reverse. In fact the 'less-developed' regions of the world are projected to account for 98 per cent of population growth over the next 25 years.[2]

In terms of regional representation, the global gene pool is obviously changing radically, but the crowds in a big-city street are a living demonstration of the fact that people from everywhere can now survive almost anywhere. The regional ecologies that determined lifestyles and moulded our species into so many ethnic groups don't carry much evolutionary weight in the new urban environment. With global trade, those who can afford it can dress according to climate, eat what they fancy from a global diet, and vaccinate or medicate against regional diseases and parasites which their bodies have never previously encountered. In modern London there are resident communities speaking 300 different languages. The kinds of ecological barriers that produced different human species in the past, and might have done so again if civilisation had not intervened, will probably remain insignificant for as long as civilisation survives. The modern constraints on migration are economic and political, but when mass movements do occur they can be very rapid.

Migrations

Humans, along with other large mammals which emerged from Africa, populated the Old World by the spread of successive generations into new climatic zones and ecological habitats. But modern humans could move in ships from one side of the world to the other in a matter of months, and with the coming of aircraft (antics of the Mile High Club notwithstanding) individuals could fly between the antipodes with no opportunity to leave any genetic trail in between. Consequently, the genetic make-up of people in different parts of the world now shows some sharp discontinuities, and it seems a fairly safe bet that, in evolutionary time, regional differences will become increasingly blurred until they finally disappear into the melting pot. While there are still plenty of social forces to slow this trend, there don't appear to be any examples of a major ethnic group actually moving in the opposite direction. In fact, some of the social and political pressures that prevent ethnic mixing in the short term can backfire spectacularly, if they contribute to economic collapse and increase the numbers of international refugees.

Intermarriage

Populations divided by colour, religion or language can retain separate identities for centuries, but that still isn't much in evolutionary time. Gradually, populations interbreed. Sometimes the mixing is surprisingly thorough and we soon forget that things were ever different. Many years ago I lived in Portugal, and my invariably pleasant memories are of a mostly stocky, Latin folk, not at all at odds with the Portuguese stereotype. And yet the genetics professor Steve Jones tells us that the Portuguese are more closely related to today's British than the British are to the Yugoslavs. He also tells us that 500 years ago the Algarve, in southern Portugal, was predominantly peopled by Africans, their genes long since absorbed into the wider population as were the African genes of eighteenth-century England.[3] Black faces are again common in the streets of Lisbon, following the strife in Portugal's former colonies of Angola and Mozambique.

The slave trade took people from West Africa to America more than 300 years ago. Today there are 30 million African-Americans but this includes many who are now of mixed race. Geneticists have calculated the genetic mixing in America in terms of a flow from white to black because the people involved prefer to call themselves 'black' (just as many Australians prefer to call themselves Aborigines when their ancestry is predominantly European). American blacks now have 20–30 per cent

European ancestry in the northern states of the USA. The figure is between 4 and 10 per cent in some southern states.[4]

International travel uncoupled genes from cultures long ago; ethnic Indians with Yorkshire accents, and Afro-Caribbean Cockneys, are now part of the fabric of British life. In the 1950s some of us firmly believed that the growing popularity of television would moderate the British regional accents and that children would all begin to sound more or less like the voices from the box. It didn't happen. Regional TV flourished and, if anything, appears to have strengthened regional differences. There is something tribal about regional identity and there is a sense of loss when languages, or even dialects, fall out of use. Societies spring up to keep them alive. We have become more ambivalent about genetic diversity; the same people who embrace the melting pot in western cities can be protesting the disappearance of remote tribes in Amazonia, just as conservationists fight to save a subspecies of bird in a patch of remnant forest. It isn't entirely illogical: there is a difference between breeding out and dying out. With the tribe it is a hunting and gathering culture that is under threat, but unlike the ubiquitous human the specialised bird with its tiny gene pool has no other way of life and nowhere else to go.

Family size

My grandmother was one of eleven children, though I think that fewer than half of them survived to maturity. That was a common scenario in those days. In contrast, I can think of several attractive young women at the moment who declare that they don't intend to have any children at all. Most people (the exceptions being the millions who still have no access to modern contraception) now have reasonable control over family size, although nobody can be certain of raising any child to maturity. Nevertheless, the direction in which we are heading points to a time when the output of future breeding stocks will have little to do with ancestral genes. If instead it becomes purely a matter of random choice, then natural selection comes to a stop and that is the end of evolution as we know it. Will it happen? Your guess, as they say, is as good as mine, but we do have a few lines of thought to follow.

LIFTING AND SHIFTING THE PRESSURES

In western societies, two-thirds of us will die from causes related to our genes, although dying after our reproductive age has passed is of no consequence to evolution.[5] For all the ethical problems that the Human Genome Project may raise, it will undoubtedly lead to an entirely new branch of medicine, with germline interventions that will affect hereditary lines in order to offer better health to children of the future—children, that is, whose parents have access to the technology. It is a tragic reality that for the immediate future, the fate of hundreds of millions of people, children included, will continue to rest largely with their innate capacity to resist the plethora of diseases, parasites and deprivations that torment them.

Human beings have a high mutation rate, and with all the new chemicals entering our systems we could be making it higher still, as well as lowering our sperm counts.[6] Diseases present another pack of wild cards. About 5 per cent of people have two copies of an allele which alters the immune system in such a way that the AIDS virus cannot enter the body's cells. This provides a natural immunity to the disease.[7] As far as we know this quirk of the body's defence mechanism was of no account 20 years ago, before the AIDS virus took off in humans, but it must now be a significant survival factor in parts of Africa where as many as one in four people are HIV positive. There could be dozens of diseases like AIDS, still unknown, and because microbes have such a rapid rate of evolution, it is a certainty that new forms will keep cropping up. The struggle for global health, for the most part, will continue to be a battle between medical science and microbial evolution in a world of changing climate.

If we imagine a world where everyone could maintain the living standards of present-day western Europe, what then? Would natural selection still be at work in the background? This is where the guesswork gets harder because we don't know whether cultural influences on reproduction would also involve genes. After all, personality types and even religiosity are heritable, so it is quite possible that those attractive young women who don't want children could have something in common genetically, while mothers who crave large families might, on average, be genetically biased in the opposite direction. If that were so, then we would be selecting genetically for people who like big families.

The truth is that nobody knows what would emerge from a few hundred generations of breeding under the circumstances that prevail in western Europe today. We can say that some selective pressures have been relaxed, because that much is obvious. Our sense of smell may be deteriorating and possibly our hearing will go the same way and we'll become more short-sighted as well. But while we insulate ourselves against selection in some directions, we might be setting up new pressures that we don't know about. The human environment is becoming predominantly urban, and some people are physically and temperamentally better suited than others to that way of life. There are huge individual differences in how well we tolerate fast foods, small spaces, rules for walking, piped music, queues, crowds, public toilets, timetables, parking problems, traffic fumes and all the other inescapables of city life. The same could be said about the pros and cons of a modern rustic existence. But is this shifting pattern of pressures doing anything to re-sort our breeding populations in terms of the types of people who, for all kinds of reasons and despite modern medicine, tend to contribute most to future generations? Is it imperceptibly changing the make-up of our gene pools such that a different distribution of physical and temperamental types would eventually exist if present trends continued? Almost certainly, would be my guess. But our present circumstances will not last: they will keep changing too. As long as humans exist, our environment will never again be as stable for anything like as long as it was during our evolutionarily formative time in the African Rift.

Ultimately it is still the climate, and dependent ecology, that creates the global setting for our stage, but the show we are putting on at present has such an enormous cast, with such a storm of technology, that much of the set already lies in ruins around us and even the climate controls may have been nudged.

ENVIRONMENTAL CHANGE AND POPULATION DENSITY

There have always been natural events that proved disastrous for the animal life involved. In the human time frame the disasters are becoming more frequent and the consequences more serious. The cost of disasters in the 1990s is said to have tripled over the previous decade.[8] Global warming, with or without human influence, is likely to bring climatic instability, and it so happens that during the past 20 years the British Isles have experienced the hottest, coldest, wettest, windiest and snowiest extremes on record, although this may only be a reflection of improved record-keeping. Putting together climatic instability and the increasing density of populations in the poorer countries, it is difficult to be optimistic about the cost of future disasters, in terms of either lives or money.

It isn't simply that the poverty-trapped people of the Third World are less able to cope with disasters than the richer communities; it is also true that populations are becoming increasingly crowded in harm's way. Millions upon millions of people now live in regions with a notoriously high risk of flooding, landslides, earthquakes or drought. In the UN's latest projections, the populations of many high-risk regions will increase by more than 40 per cent in the next 25 years. Overall, an erratic increase in the number of refugees appears to be inevitable.

Whole populations don't quietly starve to death if they can acquire land and food by fighting their neighbours, so fighting can mask ecological problems for a long time. Since the Cold War ended there have been more than 100 armed conflicts around the world, 42 of which caused over 1000 deaths a year.[9] The people displaced by such strife are rightly regarded as victims of war, but this is not to say that ecological deterioration never has anything to do with the fighting. The recent slaughter in Rwanda, for instance, was the latest outcome of centuries of hostility in that country, but we should not dismiss as irrelevant the fact that the people of Rwanda mainly live directly from their land, and with more than 300 people to the square kilometre the country is now more densely populated than Britain. Over the past 30 years the average size of a family farm in Rwanda has fallen from two hectares to 0.7 hectares.[10] Ecological problems may already feature among the many causes of armed conflict.

Not surprisingly, questions about how to handle growing pressure from migrants have generally moved up the political agendas of western governments. Speaking in Geneva in 1995, the UN High Commissioner for Refugees (UNHCR) said candidly that 'one of the most difficult problems confronting my Office in recent years has been the decline of asylum . . . Many countries are openly admitting their weariness with large numbers of refugees and blatantly closing their borders'.[11] The total population of concern to the

UNHCR remains well above 20 million and it is proving difficult to make even temporary arrangements for them. A time frame of centuries is literally unthinkable.

Global warming expands the sea and makes it rise, adding water as well if glaciers start to melt. In the last interglacial period, sea levels were 2–6 metres higher than they are now. The sea is again expanding, having risen about 10–15 centimetres during the past century.[12] Nobody can be certain what will happen in the next century or two, but a one-metre rise could have a serious impact on land up to five metres above sea level. The sort of damage that is caused by storm surges once a century, for example, could be expected about every fifteen years if the sea was a metre higher.[13] All around the world saltwater would move upstream into rivers, contaminate underground water resources and prevent or restrict agriculture on much more land than was actually flooded. But the flooding alone would be disastrous for a number of countries. The sea would cover about one-sixth of Bangladesh and a comparable proportion of Egypt's best food-producing land, where a large proportion of the population is concentrated.[14] Some studies suggest that, worldwide, a billion people would be directly affected by a sea-level rise of just one metre.[15] The costs in losses of livestock, buildings and other infrastructure are incalculable. Nor would the losses be confined to agricultural land: many coastal cities including Cairo, Calcutta, Lagos, Miami, Bangkok, New Orleans, Taipei and Venice are all perilously close to sea level. Some tiny island nations in the South Pacific would cease to exist altogether.

Mass population shifts of this magnitude, and the social upheavals that would follow all over the globe, would change the world's political geography and stir the mix of human genes in ways that are impossible to predict. The global effects, though profound, would take effect gradually as one way of life after another became uneconomical and had to be abandoned. Future historians could be recounting a very long period of painful adjustment to what would truly be a New World Order.

PEOPLE IN CONTROL

Nothing in our evolution has prepared us to think about the far-distant future. It is within our experience to think in terms of days or even decades but, although we understand the concept of thousands or millions of years, those time spans, like astronomic distances, are just measures; they convey information but not in any familiar context. Although we are thinking, planning creatures, we are basically short-term strategists. Our leaders are no different from the rest of us: the despot is concerned with retaining power from one day to the next, while the thinking of more democratic leaders is finely tuned to the length of the electoral term. And politicians of all stripes, together with big business, invariably adopt 'growth' as a high priority. True, some of them now talk about 'sustainable development' rather than growth but, for reasons discussed in Chapter 11, it usually comes down to the same thing. Big business abhors a contracting market, and I have never heard a politician sounding pleased about a diminishing constituency.

Every leader, if pressed, would have to admit that nothing can keep growing for ever, but the limits are always said to be somewhere in the future, and in somebody else's term of office. As I write this, there is a news report on my desk to say that Australia's peak housing and development body has just formed a lobby group to step up pressure for a doubling of Australia's population within the next 50 years. With the support of 'high profile business identities' the group is 'motivated largely by the desire to create more housing demand . . .'. According to the lobby group, environmental scientists who oppose this way of thinking are 'largely misguided'.[16]

I said earlier that economics had come to mediate ecology in industrialised societies, but it goes far beyond mediation. Governments now try to manipulate human reproduction and migration in order to meet the needs of economies. The economic tail tries to wag the demographic dog. More young people are being sought to fill jobs while people over 50 are lucky to be offered work once they have been made redundant. The buzzword is growth. Governments of a few nations that have passed through the demographic transition and achieved declining birth rates are concerned about the ageing of their populations. After the war there was a baby boom and, now that the babies are ageing and there are fewer youngsters to replace them, governments are favouring tax structures and other means of encouraging reproduction and another baby boom. Those babies, too, will grow old and create another 'ageing population', but that will be the problem of another government, aeons away in political terms.

The truth is that a reasonably stable population, of any secure size, has a fairly constant structure from one generation to the next, with neither booms nor busts to work through the population profile. For long periods our population was relatively stable, with just a few million of us. But that is biology. For politicians, the military and big business, there is strength in numbers. In the corporate world, stability is another word for stagnation. Growth is the buzzword.

Perhaps we all feel a strength in numbers. Perhaps it is a human trait and only natural. We know that our leaders are no less natural than the rest of us. Once again I'm stating the obvious but it hasn't always been so openly recognised, and certainly not by the leaders of the past. The leaders of old, from the high priests of city-states to the English monarchs, claimed—and quite probably believed—that they ruled by divine right and were therefore being guided by the timeless hand of the supernatural. Eternity would take care of itself if God could be kept on side. It is a way of thinking still very much alive in the modern world.

GODS IN CONTROL

Once in northern India I was being driven down a steep, winding road in a rattling old taxi when suddenly, on the opposite side, there came into view a statue of the Hindu god, Lord Ganesh. My driver turned to face it as we passed and he bowed his head and raised both hands in prayer. I grabbed the wheel in a panic but he resumed control and smiled. 'Our lives are in God's hands', he assured me. I told him I'd rather we were in the hands of a good taxi driver. From then on he seemed to think I was a comedian.

Statistics suggest that most people in the world still believe that the facts of our existence are to be found in scriptures of one sort or another and that our personal lives are part of a supernatural Grand Plan. About one in three people in the world is either a Hindu or a Muslim. About another third of the world's population is Christian, and that includes hundreds of millions who accept the literal truth of the Bible. However, it is probably true that in all westernised societies religion no longer has the influence over people's behaviour that it used to have. In some cases this is simply because Church and State no longer work in tandem. As recently as the middle of the twentieth century, as a result of religious pressure, there were states in the USA where medical practitioners could be prosecuted for giving contraceptive advice even to non-Catholics.[17] In our increasingly secular western societies, individuals are now generally free to disregard religious counsel, at least in the eyes of the law. Italians, contrary to the teaching of their Church, now practise contraception to such an extent that Italy has one of the lowest birth rates in Europe.

In November 1999, a Gallup poll was conducted by telephone across Great Britain to get some measure of the state of Christianity. Of more than 1000 people interviewed 64 per cent called themselves Christians, but well over half of them admitted going to church 'very seldom' or 'only for christenings, weddings and funerals'. As one analyst put it: 'for many, being a Christian means merely not being a Hindu, Muslim or Jew—or possibly subscribing in a vague sort of way to something called "the Christian ethic" '. Of the Christians, 71 per cent had no idea that the Millennium celebrated the birth of Christ; 34 per cent were unable to name any of the first four books of the New Testament; nearly two-thirds had no belief in the Devil and roughly one in seven self-described Christians said they didn't believe in God.

But despite all this ignorance and rejection, no fewer than 78 per cent of the Christians said they would be inclined to turn to God for help in a personal crisis, so the revelations of the Gallup poll do not mean, even in Britain, that religious feeling is going away or that people are becoming more scientific.[18] There are other features of modern societies—the fragmentation of social life, the disappearance of community and the growth of massive bureaucracies—that are thought to work against traditional religion. On the other hand it is said that we are becoming more rational, and I suppose that could be a consequence of science.[19]

The churches have not been immune to what Steve Bruce, Professor of Sociology at the University of Aberdeen, calls the erosion of the supernatural. What were once major elements of Christian faith, such as the Virgin birth and eternal damnation, are now explained away as something other than literal truths, except perhaps in parts of the world where congregations are still happy to believe in them. Religion in the West is now much more of an individual thing: a pick and mix of remnants of Church teachings and personal beliefs. It is not so much religiosity as organised religion that has shown the marked decline.

The Pope is said to have been disappointed that Poland, having escaped from communism, did not solidly return to Catholicism. Some scientists may have been similarly disappointed that so many who gave up on religious tradition did not see more value in science. Instead, during the 1970s there was a surge of western interest in all kinds of alternative religions and religious therapies such as Scientology, the Unification Church, Transcendental Meditation, Primal Therapy, Rebirthing and Bioenergetics.[20]

If the 1970s was marked by the search for more personal choice in religious thinking, then the 1980s was a time of discovery, because towards the end of that decade people were finding spiritual significance in just about anything that could be brought under the banner of the New Age. I am no expert on New Age thinking but the core of it seems to be that if something works for you, whether it be crystals, thought channelling or aromas, then it works. No matter that it doesn't work for somebody else. One could say exactly the same thing about prayer. Now, medical science is well aware of the interactions between one's state of mind and the state of one's body. Strong feelings have physiological effects, regardless of whether there are any powers in the things we feel strongly about. But New Age followers do not concern themselves with controlled experiments and statistical tests to determine cause and effect. In Bruce's summing up: 'If one takes the view that science is not a body of conclusions ... but a series of principles of discovery and testing, then it is clear that New Age science has more in common with religion than with science'.[21] And if media reports were any indication, there were many Americans who, although they dabbled in New Age behaviour themselves, were not at all pleased to learn that Ronald Reagan, when he was the world's most powerful man, was being advised on matters of state by a wife who consulted an astrologer.

With hindsight we should not be surprised that when westerners were abandoning the established churches in droves, and the cult of the individual was going from strength to strength, the two trends should have come together in the New Age. For as Bruce also points out, 'For all its talk of community, the New Age is the embodiment of individualism'.[22] And it is individualism, rather than the specifics of New Age behaviour, which seems likely to characterise western society for as long as we care to guess.

Of course, conventional religion is by no means a thing of the past in any part of the western world, nor is the conflict between biology and Christianity finally over, as several writers have claimed. In October 1996 the Pope stated that 'Fresh knowledge leads to recognition of the theory of evolution as more than just a hypothesis'.[23] That was regarded as a newsworthy sentence at the time. But if evolution, Vatican-style, has to be guided by an all-knowing deity, then science's blind forces of natural selection were not what the Pope had in mind. And now, in the third millennium, some powerful fundamentalists in the US state of Kansas have seen to it that children in state schools must learn a version of biology that has no evolution at all.[24] For the biology teachers involved, complying with this edict will be rather like having to explain the history of aviation without mentioning gravity.

It is a phenomenon worthy of social study that the Church plays a much more prominent role in America than it does in Europe these days. A survey by the Gallup Organization in 1997 found that 44 per cent of Americans believed that 'God created man pretty much in his present form at one time within the last 10 000 years'. A further 39 per cent could accept that man evolved over millions of years but believed that God guided the process. Only one in ten Americans believed the Darwinian explanation of evolution.[25] In this, the beliefs of the American public are in stark contrast to those of American senior scientists, as we shall see shortly.

In global terms: what people believe, what they are taught and what they are encouraged to do in the name of religion has such profound implications that one may wonder why biology is a particular bogey-man for so many social scientists. It is hard to say what part religion has played in recent armed conflicts around the world, but it certainly features in a fair proportion of them. Hindus fight Muslims on the Indian subcontinent, Muslims attack Christians in Indonesia, Christians attack Muslims in Yugoslavia, Sunni and Shi'ah sects of Islam squabble with each other in western Asia and Catholics fight Protestants in Northern Ireland. And that's without mentioning the Middle East. It is a situation set to last well into the future for, as sociologists point out, the world of modern humans is based on nation-states, and religious nationalism is highly compatible with the system.[26]

A FINAL THOUGHT

Among Richard Dawkins' many critics are those who say that he is on a mission to win the world for atheism—trying to make a religion out of science just as the creation scientists hoped to make a science out of religion. Certainly Dawkins thinks that we would be better off if we were free of dogma and superstition and could all think more objectively, testing our opinions against the evidence with open minds. As a professor of the public understanding of science it is his job to think like that. And he is in good company among scientists: American researchers in 1998 found that members of the National Academy of Sciences were overwhelmingly non-believers, with only 7 per cent expressing belief in a personal God.[27] The idea of a non-personal god, some unknown power beyond the natural world—deism instead of theism—is more acceptable, especially to physicists and astronomers. In any case, Dawkins must know that he preaches largely to the converted and that on a global scale his enviably broad congregation is still quite small. To my own readers I can confidently say that for every one of you who has stayed with this book to the end, there will be tens of thousands who will never hear of it.

Religion in some form or other will endure into the distant future because religious feeling, subject to that ever-present individual variation, is an inherent part of our make-up, and science will always be limited in what it can tell us. Some people can respect the scientific method, readily accept what science discovers, and yet still believe that we are all part of some Grand Plan. Many believe in an afterlife. Others can only accept the scientific limits to knowing. Some of the latter individuals, lacking religious faith, will be resigned, if not content, to recognise themselves for what they are as thinking animals, and they will marvel at it and cherish the richness of the planet that produced them. For as far as science can tell us at present, our kind is all alone in this vast universe, and when we look to the long journey of evolutionary time there is nobody at the controls.

NOTES

PREFACE
1. Frederick Seitz, 'Decline of the generalist', *Nature*, vol. 403, no. 6769, 2000, p. 483.

CHAPTER 1   KNOWING AND BELIEVING
1. David Friend with editors of *Life* Magazine, *The Meaning of Life*, Time Inc. and Little Brown, Toronto, 1991.
2. Edward B. Tylor, *Anthropology: An Introduction to the Study of Man and Civilization*, Appleton, New York, 1881, quoted in Moore, 1993 (note 5).
3. R.M. and C.H. Berndt, *The World of the First Australians*, 2nd edn, Ure Smith, Sydney, 1977, pp. 136–8.
4. Margaret Murray, *The Splendour that Was Egypt*, Sidgwick & Jackson, London, 1964, pp. 92–9. First published in 1949, this book was reprinted many times before being revised in 1964. It has been reprinted several times since then. As an introduction to life in ancient Egypt, it remains a delight to read.
5. John A. Moore, *Science as a Way of Knowing: The Foundations of Modern Biology*, Harvard University Press, Cambridge, MA, 1993, p. 30. A scholarly yet easily accessible history of biology, from the Stone Age to the present.
6. Charles Singer, *A Short History of Scientific Ideas to 1900*, Oxford University Press, London, 1959, p. 17.
7. Noah E. Fehl, *Science and Culture*, Chung Chi Publications, University of Hong Kong, 1965, pp. 169–79.
8. W.D. Ross, ed., *The Works of Aristotle*, English translation in 12 volumes, Oxford University Press, 1908–1952. Rated by *Encyclopaedia Britannica* as the best English version of Aristotle's work, it was adapted in 1952 as two books (8 & 9) in the *Britannica Great Books of the World* series. Book 8 comprises the first volume of *The Works* and includes *De anima* (on the soul). Book 9 comprises the second volume of *The Works* and

contains the biological treatises—including *Historia animalium* and *De partibus.*

9. J. Beaujeu, 'Medicine', in *Ancient and Medieval Science from Prehistory to AD 1450*, ed. R. Taton, Thames & Hudson, London, 1963, pp. 341–71. First published in French in 1957, this is one of four volumes in the series *General History of the Sciences*, all edited by R. Taton.

10. R. Arnaldez and L. Massignon, 'Arab science', in Taton, *Ancient and Medieval Science from Prehistory to AD 1450*, pp. 385–421.

11. Paul Benoît, 'Theology in the thirteenth century: a science unlike the others', in *A History of Scientific Thought*, ed. M. Serres, Blackwell, Oxford, 1995, pp. 222–45. First published in French in 1989.

12. P. Delaunay, 'Human biology and medicine', in *Beginnings of Modern Science: From AD 1450 to 1800*, p. 144 (second of the four volumes edited by R. Taton; see note 9).

13. John Henry, *The Scientific Revolution and the Origins of Modern Science*, Macmillan, London, 1997, p. 4. An excellent paperback of only 137 pages.

14. ibid., p. 74.

15. Francis Bacon, *The New Organon and Related Writings*, ed. Fulton Anderson, Bobbs Merrill, Indianapolis, 1960.

16. William C. Dampier, *A History of Science and its Relation to Philosophy and Religion*, 4th edn, Cambridge University Press, Cambridge, 1948, p. 144.

17. The reference to nailing alert dogs to dissection boards comes from a letter from a contemporary of Descartes. It was cited in Tom Regan, *All that Dwell Therein: Animal Rights and Environmental Ethics*, University of California Press, Berkeley, 1982, p. 5.

18. Derk Bodde, *Chinese Thought, Society and Science*, University of Hawaii Press, Honolulu, 1991.

19. The main elements from various dictionary definitions.

20. Martin Rees, *Before the Beginning: Our Universe and Others*, Addison Wesley, Reading, MA, 1997, p. 2.

21. There are many books on the practice and philosophy of science, some of them heavy going. Martin and Inge Goldstein, *The Experience of Science: An Interdisciplinary Approach*, Plenum, New York, 1984, is a reader-friendly, college-level

introduction to the nature of science. Carl Sagan takes a more popular approach to the scientific way of thinking in *The Demon-Haunted World: Science as a Candle in the Dark*, Headline Books, London, 1996.

22. Martha Vicinus and B. Nergaard, *Ever Yours, Florence Nightingale: Selected Letters*, Virago Press, London, 1989, p. 200.

CHAPTER 2   WHAT IS THIS THING CALLED LIFE?

1. Bob Holmes, 'Life is . . .', *New Scientist*, vol. 158, no. 2138, 1998, pp. 38–42.

2. Anon. 'Superbug survival', *New Scientist*, vol. 159, no. 2151, 1998, p. 28. See also R. John Parkes, 'A case of bacterial immortality?', *Nature*, vol. 407, no. 6806, 2000, pp. 844–5.

3. Paul Davies, *The Fifth Miracle*, Allan Lane, London, 1998, p. 61. A non-technical review of current theories on the origins of life.

4. Daniel L. Hartl, *Essential Genetics*, Jones & Bartlett, Sudbury, MA, 1996, p. 160. An undergraduate textbook.

5. Edward J. Steele, R.A. Lindley and R.V. Blanden, *Lamarck's Signature*, Allen & Unwin, Sydney, 1998, p. 226. Though intended for the general reader, this is quite a technical book. The authors believe that changes in the immune system, caused by life events, can cause genetic changes and become hereditary. The book sets out the evidence but some specialist reviewers remain unconvinced.

6. Steve Jones, *The Language of the Genes*, Flamingo, London, 1994, pp. 69–71. A prize-winning book for the general reader. Based on the 1991 Reith Lectures.

7. Georgina Ferry, 'The human worm', *New Scientist*, vol. 160, no. 2163, 1998, pp. 32–5.

8. Roger H. Reeves, 'Recounting a genetic story', *Nature*, vol. 405, no. 6784, 2000, p. 283–4.

9. Andy Coghlan and N. Boyce, 'The end of the beginning: the first draft of the human genome signals a new era for humanity', *New Scientist*, vol. 167, no. 2245, 2000, pp. 4–5.

10. Steve Jones, *The Language of the Genes*, p. 289.

CHAPTER 3   BEGINNINGS

1. Martin Rees, *Before the Beginning: Our Universe and Others*, Addison Wesley, Reading, MA, 1997, p. 1. Easy and enjoyable

reading. Includes latest theories on the origins and life cycles of the universe.

2. Stephen Hawking, *A Brief History of Time: From the Big Bang to Black Holes*, Bantam, New York, 1988, pp. 155–69.

3. Paul Davies, 'The day time began', *New Scientist*, vol. 150, no. 2027, 1996, pp. 30–5.

4. COBE stands for Cosmic Background Explorer. The finding that there were 'ripples in the fabric of space-time' was announced in 1992 by the astrophysicist George Smoot. For a splendidly readable account of the COBE project in which the significance of the findings are explained in plain language, see: George Smoot and K. Davidson, *Wrinkles in Time: The Imprint of Creation*, Little, Brown and Co., London, 1993.

5. Rees, *Before the Begining*, p. 13.

6. Norman H. Sleep *et al.* 'Annihilation of ecosystems by large asteroid impacts on the early Earth', *Nature*, vol. 342, no. 6246, 1989, pp. 139–42.

7. Phil Cohen, 'Let there be life', *New Scientist*, vol. 151, no. 2037, 1996, pp. 22–7.

8. Anon., 'In brief', *New Scientist*, vol. 158, no. 2140, 1998, p. 25.

9. Stephanie Pain, 'The intraterrestrials', *New Scientist*, vol. 157, no. 2124, 1998, pp. 28–32.

10. Michael Gross, *Life on the Edge*, Plenum Publishing, New York, 1998. First published in German by Spektrum, 1996. A rather technical work.

11. David W. Deamer, 'Boundary structures are formed by organic compounds of the Murchison carbonaceous chondrite', *Nature*, vol. 317, no. 6040, 1985, pp. 792–4.

12. Greta Schueller, 'Stuff of life', *New Scientist*, vol. 159, no. 2151, 1998, pp. 30–5.

13. Fred Hoyle and C. Wickramasinghe, *Lifecloud*, Dent, London, 1978.

14. Anon., 'Superbug survival', *New Scientist*, vol. 159, no. 2151, 1998, p. 28.

15. Christopher Chyba and C. Sagan, 'Endogenous production, exogenous delivery and impact-shock synthesis of organic molecules: an inventory for the origins of life', *Nature*, vol. 355, no. 6356, 1992, pp. 125–32.

16. Edward J. Steele, R.A. Lindley and R.V. Blanden, *Lamarck's Signature*, Allen & Unwin, Sydney, 1998, p. 56.

17. James P. Ferris, A.R. Hill, R. Liu and L. Orgel, 'Synthesis of long prebiotic oligomers on mineral surfaces', *Nature*, vol. 381, no. 6577, 1996, pp. 59–61.

CHAPTER 4   CHEMICALS TO CREATURES

1. Joanna Marchant, 'Life from the skies', *New Scientist*, vol. 167, no. 2247, 2000, pp. 4–5.
2. John Maynard-Smith and E. Szathmáry, *The Major Transitions in Evolution*, Freeman/Spektrum, Oxford, 1995, p. 102.
3. Heinrich D. Holland, 'Evidence for life on Earth more than 3850 million years ago', *Science*, vol. 275, no. 5296, 1997, pp. 38–9.
4. Mitochondria (singular, mitochondrion) are commonly barrel-shaped bodies and can be present in a cell in any number from one to thousands. They make a substance known as ATP (adenosine triphosphate), which is the universal energy carrier in all cells. An ATP molecule consists of three subunits: a sugar, a chain of three phosphates, and adenine—which is one of the bases occurring in DNA. The energy holding the outer phosphate groups is released by enzymes and used to power the cell. Green plants manufacture ATP using solar energy. Cells in animal bodies must use some of this energy to build their own ATP.
5. Christian de Duve, 'The birth of complex cells', *Scientific American*, vol. 274, no. 4, 1996, pp. 38–45. This author has much wider horizons in his book *Vital Dust: Life as a Cosmic Imperative*, Basic Books, New York, 1995.
6. Kwang W. Jeon, 'The large, free-living amoebae: wonderful cells for biological studies', *Journal of Eukaryote Microbiology*, vol. 42, no. 1, 1995, pp. 1–7. A technical paper describing the amoeba–bacteria symbiosis. An update of Professor Jeon's research appears as a book chapter in J. Seckbach, ed., *Enigmatic Microorganisms and Life in Extreme Environments*, Kluwer, Dordrecht, 1999, pp. 585–9. But it is even more technical.
7. Ten times more bacterial cells than human ones, according to Steve Jones, *Almost Like a Whale: The Origin of Species Updated*, Doubleday, London, 1999, p. 278.
8. Stuart Blackman, 'Safety in numbers', *New Scientist*, vol. 157, no. 2125, 1998, p. 15.
9. Lynn Margulis and D. Sagan, *What is Life?*, Simon &

Schuster, New York, 1995, p. 113. A richly illustrated and fascinating book, covering the origins and evolution of life.

CHAPTER 5   DARWIN'S BEST SELLER

1. A summary of evolutionary thinking before Darwin is included in Michael White and J. Gribbin, *Darwin: A Life in Science*, Dutton, New York, 1995. This is one of the best Darwin biographies.

2. The lines of verse are quoted by J.W. Burrow in his editor's introduction to Darwin's *The Origin of Species by Natural Selection*, Penguin, London, p. 27. This printing of the first edition was produced as a 'Penguin Classic' paperback in 1985 and is still in print.

3. ibid., p. 14.

4. John A. Moore, *Science as a Way of Knowing: The Foundations of Modern Biology*, Harvard University Press, Cambridge, MA, 1993, p. 120.

5. Charles Darwin, *The Structure and Distribution of Coral Reefs*, 1842. An edited version of this, and other major works by Darwin, can be found in Mark Ridley (ed.), *The Essential Darwin*, Allen & Unwin, London, 1987.

6. Charles Darwin, *The Origin of Species by Means of Natural Selection*, John Murray, London, 1859. See note 2.

7. Howard E. Gruber and P.H. Barrett, *Darwin on Man: A Psychological Study of Scientific Creativity*, E.P. Dutton, New York, 1974, p. 163. The authors are a psychologist and a biologist, respectively. They explore many other aspects of Darwin's thinking.

8. V. Orel, *Gregor Mendel: The First Geneticist*, Oxford University Press, Oxford, 1996. An authoritative book by the head of the Mendelianum at Brno.

9. Some writers think that Mendel's ratios came out a bit too perfectly for the data to be beyond suspicion.

10. Ernst Mayr, *One Long Argument: Charles Darwin and the Genesis of Modern Evolutionary Thought*, Penguin, London, 1991. Mayr, as one of the leading authors of the modern synthesis, offers a first-hand account of evolutionary thinking in the twentieth century.

11. Julian S. Huxley, *Evolution: The Modern Synthesis*, Allen & Unwin, London, 1942.

12. The passage is based on a report from the *New York Times*, 21 June, 1994, cited in Anthony Giddens, *Sociology*, 3rd edn, Polity Press, Cambridge, in association with Blackwell, Oxford, 1997, p. 99.

CHAPTER 6   DARWINIANS WITH COMPUTERS

1. Most of Dr Schwaner's publications are in specialist journals but the following book chapter provides an initial overview: 'Population structure of black tiger snakes, *Notechis ater niger*, on offshore islands of South Australia', in *Biology of Australasian Frogs and Reptiles*, eds G. Grigg, R. Shine and H. Ehmann, Royal Zoological Society of New South Wales, Sydney, 1985, pp. 35–46.

2. Random change in gene frequencies, known as genetic drift, can be rapid in small, isolated populations because these are likely to contain only a sample of the full gene pool to begin with. In island tiger snakes, for example, variations in venom proteins have been detected that are best explained by genetic drift.

3. The theoretical basis for speciation without isolation was recently advanced by two computer models published in, *Nature*. A short appraisal can be found in Tom Treganza and R.K. Butlin, 'Speciation without isolation', *Nature*, vol. 400, no. 6742, 1999, pp. 311–12.

4. It is possible, however, for selection to work on responses to cues that come in advance. If significant weather patterns were preceded by minor events, for example, then selection could operate on an organism's responses to those signals.

5. R.J. Berry, *Neo-Darwinism*, Studies in Biology no. 144, Edward Arnold, London, 1982, p. 33. Like all the books in the series, this is a slim paperback but packed with information.

6. Dan Nilsson and S. Pelger, 'A pessimistic estimate of the time required for an eye to evolve', *Proceedings of the Royal Society of London*, vol. B256, 1994, pp. 53–8. The paper also appears in *Evolution*, ed. Mark Ridley, Oxford University Press, Oxford, 1997, pp. 293–301. This paperback is an Oxford Reader, not to be confused with Ridley's textbook, which is also called *Evolution*.

7. Richard Dawkins, *The Blind Watchmaker*, Longman, London, 1986.

8. Richard Dawkins, *The Selfish Gene*, Oxford University Press, 1976 (2nd edn 1989). The theoretical basis for this popular book is contained in George C. Williams, *Adaptation and Natural Selection*, Princeton University Press, Princeton, NJ, 1966.

9. Stephen J. Gould's many collections of essays include: *The Panda's Thumb*, 1980; *Hens' Teeth and Horses' Toes*, 1983; *The Flamingo's Smile*, 1985; *Bully for Brontosaurus*, 1991; *Eight Little Piggies*, 1993; *Dinosaur in a Haystack*, 1996. All published by Norton, New York.

10. According to Niles Eldredge, the expression was coined by John Turner, an British geneticist.

11. Niles Eldredge, *Reinventing Darwin: The Great Evolutionary Debate*, John Wiley, New York, 1995, p. 97.

12. Stephen J. Gould and R. Lewontin, 'The spandrels of San Marco and the Panglossian paradigm: a critique of the adaptationist programme', *Proceedings of the Royal Society of London*, vol. B205, 1979, pp. 581–98. The paper is reprinted in *Evolution*, ed. Mark Ridley, Oxford University Press, Oxford, 1997, pp. 139–54.

13. The biggest 'Just So' story of all is the aquatic ape theory, first suggested in 1960 by the marine biologist Alister Hardy and later elaborated by the writer Elaine Morgan. Humans have unexplained traits, such as the diving reflex and naked skin of newborns, which might be understandable if our ancestors lived in water. After weighing all the evidence, a team of authors concluded that the arguments favouring the theory 'are not sufficiently convincing to counteract the arguments against it'. M. Roede *et al.*, *The Aquatic Ape: Fact or Fiction?*, Souvenir Press, London, 1991.

14. Daniel C. Dennet, *Darwin's Dangerous Idea*, Simon & Schuster, New York, 1995, p. 261.

15. Eldredge, in *Reinventing Darwin*, pp. 101–2, acknowledges what Gould said, but denies that Gould is a Marxist and points out that there is a difference between learning Marxism and adopting it.

16. Dennet, *Darwin's Dangerous Idea*, p. 309. ACLU is the acronym for the American Civil Liberties Union.

17. Gould, *The Panda's Thumb*, p. 90–2.

18. Mark Ridley, *Evolution*, 2nd edn, Blackwell, Oxford, 1996.

19. V.C. Wynne-Edwards, *Animal Dispersion in Relation to Social Behaviour*, Oliver & Boyd, Edinburgh, 1962. Persisting with the idea, Wynne-Edwards also wrote *Evolution Through Group Selection*, Blackwell, Oxford, 1986.

20. William D. Hamilton, 'The evolution of altruistic behaviour', *American Naturalist*, no. 97, 1963, pp. 354–6. A reprint of this short, provisional paper appears in T.H. Clutton-Brock & P. Harvey (eds) *Readings in Sociobiology*, Freeman & Co., San Francisco, 1978, pp. 31–3. A more detailed and mathematical treatment by Hamilton was published in 1964. Hamilton died in March 2000 from malarial complications following fieldwork in the Congo. He was 63.

21. Andrew F.G. Bourke, 'Sociality and kin selection in insects', in *Behavioural Ecology: An Evolutionary Approach*, 4th edn, eds J.R. Krebs & N.B. Davies, Blackwell Science, Oxford, 1997, pp. 203–27.

22. Stephen T. Emlen, 'Predicting family dynamics in social vertebrates', in *Behavioural Ecology: An Evolutionary Approach*, 4th edn, eds J.R. Krebs & N.B. Davies, Blackwell Science, Oxford, 1997, pp. 228–53.

CHAPTER 7   SHAPING OUR FAMILY TREE

1. When molten rock, such as lava, cools and crystallises, the iron it contains becomes magnetised parallel to the Earth's magnetic field.

2. Eldridge M. Moores and R.J. Twiss, *Tectonics*, W.H. Freeman, New York, 1995, p. 302.

3. For an illustrated overview see Ron Redfern, *Origins: Evolution of Continents, Oceans and Life*, Cassell, London, 2000.

4. Jared Diamond, *The Rise and Fall of The Third Chimpanzee*, Radius, London, 1991.

5. An overview of these early types is given by Elwyn Simons, 'The fossil history of primates', in *The Cambridge Encyclopedia of Human Evolution*, eds. S. Jones, R. Martin and D. Pilbeam, Cambridge University Press, Cambridge, 1992, pp. 199–208.

6. Some authorities put the robust australopithecines in a separate genus—*Paranthropus*.

7. Berhane Asfaw *et al.*, '*Australopithecus garhi*: a new species of early hominid from Ethiopia', *Science*, vol. 284, no. 5414, 1999, p. 629–35. In the same issue, Elizabeth Culotta comments on

this article in 'A new human ancestor?', pp. 572–3.

8. A summary of the australopithecines is given by Bernard Wood, 'Evolution of australopithecines' in *The Cambridge Encyclopedia of Human Evolution*, Cambridge, 1992, pp. 231–40. The Laetoli footprints were the subject of a *National Geographic* feature, 'Footprints in the ashes of time', April 1979, pp. 446–57.

9. Bernard Wood and M. Collard, 'The human genus', *Science*, vol. 284, no. 5411, 1999, pp. 65–71.

10. The tools were crude choppers and flakes that were made in Africa between about 2.5 and one million years ago. The style is known as the Oldowan tradition, after the Olduvai Gorge in Tanzania.

11. Christopher G. Janus with W. Brashler, *The Search for Peking Man*, Macmillan, New York, 1975. The story of how the fossils came to be lost and the fruitless attempts to find them.

12. Ian Tattersall, *Becoming Human*, Harcourt Brace, New York, 1998, pp. 136–40. Also Michael Balter and A. Gibbons, 'A glimpse of humans' first journey out of Africa', *Science*, vol. 288, no. 5468, 2000, pp. 948–50.

13. Rebecca L. Cann, M. Stoneking and A.C. Wilson, 'Mitochondrial DNA and human evolution', *Nature*, vol. 325, no. 6099, 1987, pp. 31–6.

14. Richard Dawkins, *River out of Eden*, Weidenfeld & Nicolson, 1995, pp. 55–66. See also Philip Cohen, 'Eve came first', *New Scientist*, vol. 168, no. 2263, 200, p. 16.

15. Evelyn Strauss, 'Can mitochondrial clocks keep time?', *Science*, vol. 283, no. 5407, 1999, pp. 1435–7.

16. The African replacement and regional continuity models are both spelled out in 'Where did modern humans originate?', *Scientific American*, vol. 266, no. 4, 1992, pp. 20–33.

17. Elizabeth Pennisi, 'Genetic study shakes up out of Africa theory', *Science*, vol. 283, no. 5409, 1999, p. 1828.

18. Günter Bräuer, 'Africa's place in the evolution of *Homo sapiens*', in *Continuity or Replacement: Controversies in Homo sapiens Evolution*, ed. Günter Bräuer, A.A. Balkema, Rotterdam, 1992, pp. 83–98.

19. Steven Rose, R.C. Lewontin and L.J. Kamin, *Not in Our Genes; Biology, Ideology and Human Nature*, Pantheon Books, New York, 1984, p. 126.

20. Tim Beardsley, 'Mutations galore', *Scientific American*, vol. 280, no. 4, 1999, p. 24.

21. Montgomery Slatkin, 'Gene flow and the geographic structure of natural populations', *Science*, vol. 236, no. 4803, 1987, pp. 787–92.

22. M.J. Morwood *et al.*, 'Fission-track ages of stone tools and fossils on the east Indonesian island of Flores', *Nature*, vol. 392, no. 6672, 1998, pp. 173–6.

23. In modern German the spelling is *Tal*, so one sometimes sees 'Neandertal'. German pronunciation is always with the hard 't'.

24. James Shreeve, *The Neandertal Enigma*, William Morrow, New York, 1995, pp. 30–4.

25. Ian Tattersall, *Becoming Human*, Harcourt Brace, New York, 1998, pp. 159 & 165.

26. Patricia Kahn and A. Gibbons, 'DNA from an extinct human', *Science*, vol. 277, no. 5323, 1997, pp. 176–8.

27. Eric Trinkaus and C. Duarte, 'The hybrid child from Portugal', *Scientific American*, vol. 282, no. 4, 2000, p. 82.

CHAPTER 8   WERE PEOPLE EVER WILDLIFE?

1. Neil Roberts, 'Climatic changes in the past', in *The Cambridge Encyclopedia of Human Evolution*, eds S. Jones, R. Martin and D. Pilbeam, Cambridge University Press, Cambridge, 1992, pp. 174–8.

2. R.H.V. Bell, 'A grazing ecosystem in the Serengeti', *Scientific American*, vol. 225, no. 1, 1971, pp. 86–93.

3. Bernard A. Wood, 'Evolution of australopithecines', in *The Cambridge Encyclopedia of Human Evolution*, pp. 231–40.

4. Gretchen Vogel, 'Did early African hominids eat meat?', *Science*, vol. 283, no. 5400, 1999, p. 303.

5. Robert Foley, *Humans before Humanity*, Blackwell, Oxford, 1995, p. 140–1.

6. Richard Potts, 'The hominid way of life', in *The Cambridge Encyclopedia of Human Evolution*, pp. 325–34.

7. Roger Lewin, 'The great brain race', *New Scientist*, vol. 136, no. 1850, 1992. This is an 8-page special feature inserted between pp. 28 and 29.

8. For critical discussion of the EQ concept see Paul H. Harvey and J.R. Krebs, 'Comparing brains', *Science*, vol. 249, no. 4965, 1990, pp. 140–6.

9. Because body weight is so variable, and because it cannot be measured in extinct species, some authorities have scaled brain size against the size of the eye orbit in the skull. The orbital area and the cranial capacity can both be measured in fossil skulls and the relationship between eye size and body size in large, diurnal primates is probably as constant as any other relationship that might be used. The study by Bernard Wood and Mark Collard, referenced in Chapter 7, was based on this method.

10. Colin Tudge, *The Day Before Yesterday: Five Million Years of Human History*, Random House, London, 1995, p. 208.

11. Robert Foley, *Humans before Humanity*, p. 171.

12. Recent articles on culture in chimps and monkeys include: A. Whiten *et al.*, 'Cultures in chimpanzees', *Nature*, vol. 399, no. 6737, 1999, pp. 682–5. And, in the same issue: Frans B.M. de Waal, 'Cultural primatology comes of age', pp. 365–6. A special issue on evolution in *Science* included: Gretchen Vogel, 'Chimps in the wild show stirrings of culture', *Science*, vol. 284, no. 5423, 1999, pp. 2070–3

13. Toshida Nishida, 'Local traditions and cultural transmission', in *Primate Societies*, eds B.B. Smuts *et al.*, 1987, pp. 462–74. *Primate Societies* is a well-illustrated and large-format paperback. It deals widely with the subject in a scientific but quite readable way.

14. Marc D. Hauser, 'Invention and social transmission: new data from wild vervet monkeys', in *Machiavellian Intelligence*, eds R. Byrne and A. Whiten, Clarendon Press, Oxford, 1988, pp. 327–43. This book is also a large-format paperback but all chapters are concerned with social intelligence. A second edition, with new contributions, was published in 1997.

15. Dan Charles *et al.*, 'Chimp learns Stone-Age use of tools', *New Scientist*, vol. 129, no. 1757, 1991, p. 21.

16. Ian Tattersall, *Becoming Human*, Harcourt Brace & Co., New York, p. 56.

17. Adrienne L. Zihlman and N.M. Tanner, 'Becoming human: putting women in evolution', paper presented at the annual meeting of the American Anthropological Society, Mexico City, 1974.

18. Nancy M. Tanner, *On Becoming Human*, Cambridge University Press, New York, 1981.

19. Reported in the Papua New Guinea daily newspaper, *The Post Courier*, August 23, 1978, p. 3.
20. Robin McKie, 'The people eaters', *New Scientist*, vol. 157, no. 2125, 1998, pp. 43–6.
21. Phylis C. Lee, 'Testing the intelligence of apes', in *The Cambridge Encyclopedia of Human Evolution*, p. 111.
22. Frans de Waal, 'Chimpanzee politics', in *Machiavellian Intelligence*, pp. 122–31. This is an extract from de Waal's book, *Chimpanzee Politics*, Jonathon Cape, London, 1982. For a more recent discussion see Laura Spinney, 'Liar! Liar!', *New Scientist*, vol. 157, no. 2121, 1998, pp. 22–6.
23. Richard D. Alexander, *Darwinism and Human Affairs*, University of Washington Press, Seattle, 1979.
24. A discussion of Theory of Mind in chimps appears in D.J. Povinelli and L.R. Godfrey, 'The chimpanzee's mind: how noble in reason? How absent of ethics?', in *Evolutionary Ethics*, eds M.H. Nitecki and D.V. Nitecki, State University of New York Press, New York, 1993, pp. 277–324.
25. Kelly J. Stewart and A.H. Harcourt, 'Gorillas: variation in female relationships', in Smuts *et al.*, *Primate Societies*, pp. 155–64.
26. David Barash, *Sociobiology and Behaviour*, 2nd edn, Hodder & Stoughton, London, 1982, p. 277. Barash is citing earlier data from G.P. Murdock, *Ethnographic Atlas*, University of Pittsburgh Press, Pittsburgh, PA, 1967.
27. Kathleen Stern and M.K. McClintock, 'Regulation of ovulation by human pheromones', *Nature*, vol. 392, no. 6672, 1998, pp. 177–9. Strictly a scientific paper. The significance of the research is discussed by A. Weller on pp. 126–7 of the same issue.
28. Paul G. Bahn and J. Vertut, *Images of the Ice Age*, Facts on File, New York, 1988. In coffee-table format, this book deals with both cave art and portable artefacts. Bahn also has a brief chapter on 'Ancient art' in *The Cambridge Encyclopedia of Human Evolution*, pp. 361–4.
29. Tattersall, *Becoming Human*, p. 27.

CHAPTER 9   ALTERING THE LANDSCAPE

1. David M. Gates, *Energy and Ecology*, Sinauer Associates, Sunderland, MA, 1985, p. 87.

2. Michael Begon, J.L. Harper and C.R. Townsend, *Ecology*, 2nd edn, Blackwell Scientific, Oxford, 1990, pp. 80–9.

3. ibid., pp. 805–6. For recent research see David M. Post, M.L. Pace and N.G. Hairston, 'Ecosystem size determines food-chain length in lakes', *Nature*, vol. 405, no. 6790, 2000, pp. 1047–9.

4. Begon, Harper and Townsend, *Ecology*, p. 652.

5. William T. Vickers, 'Hunting yields and game composition over ten years in an Amazon Indian territory', in *Neotropical Wildlife Use and Conservation*, eds J.G. Robinson and K.H. Redford, University of Chicago Press, Chicago, 1991, pp. 53–81.

6. G.B. Silberbauer, 'The G/Wi Bushmen', in *Hunters and Gatherers Today*, ed. M.G. Bicchieri, Holt, Rinehart & Winston, New York, 1972, pp. 271–326.

7. R.L. Kirk, *Aboriginal Man Adapting: The Human Biology of Australian Aborigines*, Clarendon Press, Oxford, 1981, p. 41.

8. ibid., p. 86.

9. ibid., p. 63.

10. M. Alvard, 'Testing the ecologically noble savage hypothesis: interspecific prey choice by Piro hunters of Amazonian Peru', *Human Ecology*, vol. 21, no. 4, 1993, pp. 355–87.

11. Nathan Keyfitz, 'The growing human population', *Scientific American*, vol. 261, no. 3, 1989, pp. 70–7.

12. Ian Tattersall, *Becoming Human*, Harcourt Brace & Co, New York, 1998, p. 206.

13. Peter Atkins, I. Simmons and B. Roberts, *People, Land and Time*, Arnold, London, 1998, p. 16. The same point is made by D.R. Harris, 'Human diet and subsistence', in *The Cambridge Encyclopedia of Human Evolution*, Cambridge University Press, Cambridge, 1992, pp. 69–74.

14. I.G. Simmons, *Changing the Face of the Earth*, 2nd edn, Blackwell, Oxford, 1996, p. 60. An outstanding textbook on human geography.

15. Douglas T. Price and A.B. Gebauer (eds), *Last Hunters First Farmers*, School of American Research Press, Santa Fe, 1995. This book was the outcome of a seminar on the origins of agriculture, and there are many references to the influence of climate.

16. Barbara Bender, *Farming in Prehistory*, Faber, London, 1975, p. 5.

17. L. Palmqvist, 'The great transition: first farmers of the western world', in *People of the Stone Age*, ed. Göran Burenhult, Harper Collins, San Francisco, 1993, pp. 16–21. This is Volume 2 in *The Illustrated History of Humankind*; an authoritative but richly illustrated and non-technical series.

18. Jared Diamond, *Guns, Germs and Steel*, Jonathon Cape, London, 1997, p. 100. The author, who is Professor of Physiology at the University of California, explains some of the problems involved in carbon dating. With the latest techniques, samples as small as a single seed can be dated.

19. Carles Vila *et al.*, 'Multiple and ancient origins of the domestic dog', *Science*, vol. 276, no. 5319, 1997, pp. 1687–9. There were nine authors in this international research team. The findings were summarised by Bob Holmes in *New Scientist*, vol. 154, no. 2087, 1997, p. 19.

20. Ofer Bar-Yosef and R.H. Meadow, 'The origins of agriculture in the near east', in Price and Gebauer, *Last Hunters First Farmers*, pp. 39–94.

21. Peter Rowley-Conwy, 'Abu Hureyra: the world's first farmers', in Burenhult, *People of the Stone Age*, pp. 27–9.

22. Gary O. Rollefson, 'Ain Ghazal: the largest known Neolithic site', in Burenhult, *People of the Stone Age*, pp. 36–7.

23. Douglas T. Price, A.B. Gebauer and L.H. Keeley, 'The spread of farming into Europe north of the Alps', in Price and Gebauer, *Last Hunters First Farmers*, pp. 95–126.

24. Palmqvist, 'The great transition', in Burenhult, *People of the Stone Age*, pp. 36–7.

25. Charles L. Redman, 'Mesopotamia and the first cities: 4000–539 BC', in *Old World Civilizations*, ed. Göran Burenhult, Harper Collins, San Francisco, 1994, pp. 17–37. Volume 3 in *The Illustrated History of Humankind*.

26. ibid.

27. Simmons, *Changing the Face of the Earth*, p. 27.

28. ibid., p. 64.

29. ibid., p. 143.

30. Norman Kretchmer, 'Lactose and lactase', *Scientific American*, vol. 227, no. 4, 1972, pp. 70–8.

31. The figures are for non-identical twins. Identical twins occur in about 3–5 pregnancies per 1000 in all populations. Arthur P. Mange and E.J. Mange, *Genetics: Human Aspects*, 2nd edn,

Sinauer Associates, Sunderland, MA, 1990, p. 466.

32. Diamond, *Guns, Germs and Steel*, p. 207.

33. Kirk, *Aboriginal Man Adapting*, p. 172.

34. Feng Gau *et al.*, 'Origin of HIV-1 in the chimpanzee *Pan troglodytes troglodytes*', *Nature*, vol. 397, no. 6718, 1999, pp. 436–40. A strictly scientific research paper authored by a dozen scientists.

CHAPTER 10   DARWIN, GOD AND SOCIETY

1. Kingsley Davis, 'Urbanization of the human population', in *Readings from Scientific American*, W.H. Freeman, San Francisco, 1972, pp. 373–85.

2. Robert Nisbet, *The Social Philosophers*, Heinemann Educational, Oxford, 1974, p. 152.

3. Thomas Hobbes, *Leviathan*, Penguin, London. This Pelican Classic was published in 1968 from the original work of 1561. It runs to more than 700 pages. The quotes are from Ch. 17, 'Of Common-wealth', pp. 226–7.

4. ibid., Ch. 13, 'Of Man', p. 186.

5. ibid., Ch. 29, 'Of Common-wealth', pp. 370–1.

6. Nisbet, *Social Philosophers*, p. 153.

7. 'When wild in woods the noble savage ran', John Dryden, *The conquest of Granada by the Spaniards*, pt. i, I. i, 1672.

8. An extract from Hume's *Enquiry Concerning Human Understanding*, in Leslie Stevenson, *The Study of Human Nature*, Oxford University Press, Oxford, 1981, p. 118.

9. Hume's dialogue on *The Argument from Design* appears in *Evolution*, ed. Mark Ridley, Oxford University Press, Oxford, 1997, pp. 387–9.

10. Richard Dawkins' book, *The Blind Watchmaker*, was written with this very question in mind.

11. This remark comes from Beecher's *Evolution and Religion*, 1885, quoted in Richard Hofstadter, *Social Darwinism in American Thought*, Beacon Press, Boston, 1955, p. 29.

12. Details on Spencer are from Hofstadter, *Social Darwinism in American Thought*, pp. 31–50.

13. ibid., pp. 51–66, for details on Sumner, including quotations.

14. Steve Jones, *The Language of the Genes*, Flamingo, London, 1994, p. 2.

15. Hofstadter, *Social Darwinism*, p. 162.

16. Francisco J. Ayala, *Population and Evolutionary Genetics: A Primer*, Benjamin Cummings, Menlo Park, CA, 1982, p. 99.
17. Jones, *Language of the Genes*, p. 12.
18. Nisbet, *Social Philosophers*, p. 379.
19. Mentioned by J.W. Burrow, in his 'Editor's introduction' to *The Origin*, Penguin, 1985, p. 45.
20. Hofstadter, *Social Darwinism*, p. 115.
21. Julian Huxley, *Soviet Genetics and World Science*, Chatto & Windus, London, 1949, p. 26.
22. Lysenko coined the term 'vernalisation' (*yarovizatsia* in Russian) for a technique that had been discovered decades earlier in Germany, and was possibly known before that.
23. Dominique Lecourt, *Proletarian Science? The Case of Lysenko*, NLB, London, 1977, pp. 42–3. In fact, it is possible for the action of genes to be influenced by environmental shock, though this 'epigenetic inheritance' has not yet been confirmed in mammals under natural conditions. For a non-technical summary see Gail Vines, 'Hidden inheritance', *New Scientist*, vol. 160, no. 2162, 1998, p. 27–30. Also Michael Balter, 'Was Lamarck just a little bit right?', *Science*, vol. 288, no. 5463, 2000, p. 38.
24. Huxley, *Soviet Genetics*, pp. 29–30.
25. ibid., p. 38.
26. Lecourt, *Proletarian Science*, p. 27.
27. ibid., p. 18.
28. ibid., pp. 132–3.
29. Edward O. Wilson, *On Human Nature*, Harvard University Press, Cambridge, MA, 1978, p. 21.
30. Philip K. Bock, *Modern Cultural Anthropology*, Knopf, New York, 1974, pp. 430–4.
31. The prizewinners were Karl von Frisch, best known for his work on bees; Konrad Lorenz, who was a pioneer of the science of ethology; and Nikolaas Tinbergen, who organised a research department in ethology at Oxford University.
32. Edward O. Wilson, *Sociobiology: The New Synthesis*, Harvard University Press, Cambridge, MA, 1975. An abridged, paperback edition of 366 pages was produced by the same publisher in 1980.
33. Daniel G. Bates and F. Plog, *Cultural Anthropology*, Allyn & Bacon, Needham Heights, MA, 1996, p. 22.

34. Edward Wilson has expressed his 'dream of unified learning' in his latest book: *Consilience: The Unity of Knowledge*, Little, Brown & Co., London, 1998.

35. Ullica Segerstråle, *Defenders of the Truth: The Battle for Science in the Sociobiology Debate and Beyond*, Oxford University Press, Oxford, 2000.

CHAPTER 11   EXPLOSIVE TIMES: HUMAN ECOLOGY IN
THE TWENTIETH CENTURY

1. Ian Tattersall, *Becoming Human*, Harcourt Brace & Co., New York, p. 161.

2. R. S. Meindl, 'Human populations before agriculture', in *The Cambridge Encyclopedia of Human Evolution*, pp. 406–10.

3. Kate Douglas, 'Making friends with death-wish genes', *New Scientist*, vol. 143, no. 1936, 1994, pp. 31–4.

4. To watch it happen on a calculator, start with 0.0001 metre—which is about the thickness of an ordinary sheet of typing paper.

5. Julian Huxley, *Essays of a Humanist*, Chatto & Windus, London, 1964, p. 246.

6. Pramilla Senanayake, 'Women and the family planning imperative', in *Population and Global Security*, ed. N. Polunin, Cambridge University Press, Cambridge, 1998, pp. 185–204.

7. Vaclav Smil, *Global Ecology: Environmental Change and Social Flexibility*, Routledge, London, 1993, p. 47.

8. FAO, *Dimensions of Need: An Atlas of Food and Agriculture*, UN Food & Agriculture Organization, Rome, 1995, p. 21.

9. UNFPA, *The State of World Population 1999*, UNFPA, New York, 1999, p. 28. An annual report by the UN Population Fund which includes current demographic data.

10. Lester Brown, 'Facing the prospect of food scarcity', in *State of the World 1997*, eds L. Brown *et al.*, Earthscan Publications, London, 1997, pp. 23–41. Annual reports of the Worldwatch Institute, of which this is the 14th edition, tend to be rather doom-laden but are well-referenced and include useful data from official sources.

11. FAO, *Dimensions of Need*, p. 14.

12. Brown, 'Facing the prospect', p. 35.

13. Gary Gardner, 'Preserving global cropland', in *State of the World 1997*, eds L. Brown *et al.*, Earthscan Publications, London, 1997, pp. 42–59.

14. A.J. McMichael, *Planetary Overload: Global Environmental Change and the Health of the Human Species*, Cambridge University Press, Cambridge, 1993, pp. 88–9.
15. Brown, 'Facing the prospect', p. 35.
16. FAO, *Dimensions of Need*, p. 53.
17. Debora Mackenzie, 'The cod that disappeared', *New Scientist*, vol. 147, no. 1995, 1995, pp. 24–9.
18. *World Resources 1998–1999*, a joint publication of the World Resources Institute, the United Nations Development Program and the World Bank, Oxford University Press, New York, 1998, p. 159.
19. UNFPA, *The State of World Population 1999*, p. 3.
20. FAO, *Dimensions of Need*, p. 36.
21. *World Resources 1898–1999*, p. 147.
22. UNFPA, *The State of World Population 1999*, pp. 70–2. Estimates of per-capita energy consumption are given in kilograms of oil equivalent. Conversions were based on 1 koe = 10 000 kcal.
23. John H. Gibbons, P.D. Blair and H. Gwin, 'Strategies for energy use', *Scientific American*, vol. 261, no. 3, 1989, pp. 86–97. This issue of the journal was devoted to articles on 'Managing planet earth' and was published as a paperback with that title by W.H. Freeman, New York, 1990.
24. L.R. Taylor (ed.), *The Optimum Population for Britain*, Symposia of the Institute of Biology, no. 19, London, 1969, p. 165.
25. Laurie Garrett, *The Coming Plague*, Penguin, New York, 1994, p. 252.
26. WCED, *Our Common Future*, Report of the World Commission on Environment and Development, Oxford University Press, Oxford, 1987, pp. 28–31.
27. P.K. Gupta, *Pesticides in the Indian Environment*, Interprint, New Delhi, 1986.
28. On a per-capita basis, the people of several income-poor countries, including Malaysia, Ecuador and Madagascar, are estimated to be producing more CO_2 gas by burning forest and grass than are the people of industrial nations, such as France and Japan, by burning fossil fuels (Smil, *Global Ecology*, p. 187).
29. This equation has been elaborated and discussed by many

authors over the years but it had its origins in an article by P.R. Ehrlich and J.P. Holdren, 'Impact of population growth', *Science*, vol. 171, no. 3977, 1971, pp. 1212–7.

30. Details are published by the Murray–Darling Basin Commission, Canberra, and are posted on their website: http://www.mdbc.gov.au.

31. Sandra Postel, 'Saving water for agriculture', in *State of the World 1990*, eds L. Brown *et al.*, Unwin Hyman, London, 1990, p. 45.

32. Smil, *Global Ecology*, p. 48.

33. Brown, 'Facing the prospect', p. 30.

34. UNFPA, *The State of World Population 1999*, p. 28.

35. Brown, 'Facing the prospect', pp. 29–30.

36. WCMC, *Global Biodiversity: Status of the Earth's Living Resources*, compiled by World Conservation Monitoring Centre, Chapman & Hall, London, 1992, p. xiii. A useful source of facts and figures.

37. Julian L. Simon and A. Wildavsky, 'On species loss, the absence of data, and risks to humanity', in *The Resourceful Earth: A Response to Global 2000*, eds J. Simon and H. Kahn, Blackwell, Oxford, 1984, pp. 171–83.

38. *World Birdwatch*, Journal of BirdLife International, vol. 21, no. 3, Oct. 1999, pp. 9–12.

39. M. Bolton (ed.), *Conservation and the Use of Wildlife Resources*, Chapman & Hall, London, 1997.

40. Scott Norris, 'Strictly for the birds', *New Scientist*, vol. 166, no. 2243, 2000, p. 11. A brief report on research from the annual conference of the Society for Conservation Biology.

41. WCMC, *Global Biodiversity*, p. 460, with updates from other sources.

42. WCED, *Our Common Future*, pp. 165–6.

43. M. Bolton, 'Ecodevelopment in the crowded tropics: what prospects for conservation?', *Environmental Conservation*, vol. 21, no. 3, 1994, pp. 259–62.

44. Jane Seymour, 'Freezing time at the zoo', *New Scientist*, vol. 141, no. 1910, 1994, pp. 21–3.

45. UNFPA, *The State of World Population 1999*, p. 19.

46. D.E.C. Eversley, 'The special case—managing human population growth', in *The Optimum Population for Britain*, ed. L.R. Taylor, Symposia of the Institute of Biology, no. 19, London,

1969, pp. 103–16. In another chapter of this volume, a contributor from the General Register Office confirms the 1930s forecast of a static or declining population for England and Wales.

47. A.J. McMichael, *Planetary Overload: Global Environmental Change and the Health of the Human Species*, Cambridge University Press, Cambridge, 1993, p. 247.

48. FAO *Dimensions of Need*, pp. 82–3.

49. Frank Furedi, *Population and Development: A Critical Introduction*, Polity Press, Cambridge, 1997, pp. 142–61.

CHAPTER 12   RECIPES AND OUTCOMES

1. The biggest commercial venture involved, Craig Venter's Celera Genomics, became fully independent in 1998. Celera now rivals the publicly funded project in technical output. Up-to-date information on the HGP is posted on the websites (base URL www.ornl.gov/hgmis) maintained by the US Department of Energy.

2. Philip Cohen, in an eight-page special report on the Genome Project, *New Scientist*, vol. 166, no. 2239, 2000, p. 19.

3. This is standard textbook material. Further details can be found in Peter H. Raven and G.B. Johnson, *Biology*, 4th edn, W.C. Brown, Dubuque, IA, 1996, pp. 349–69.

4. Francisco Ayala, *Population and Evolutionary Genetics*, Bejamin Cummings, Menlo Park, CA, 1982, p. 172.

5. Richard Dawkins, *River out of Eden*, Weidenfeld & Nicholson, London, 1995, p. 95.

6. Nancy L. Segal, *Entwined Lives: Twins and What They Tell Us About Human Behaviour*, Dutton, New York, 1999, p. 295.

7. Each child gets a different mix of 50 per cent of the genes of each parent, though Steve Jones tells us that there have been cases of fraternal twins having had different fathers (*Language of the Genes*, HarperCollins, London, 1994, p. 235).

8. Steven Rose, R.C. Lewontin and L.J. Kamin, *Not in Our Genes*, Pantheon, New York, 1984, p. 103. Also published by Penguin, London, in 1990 and still in print.

9. Leslie S. Hearnshaw, *Cyril Burt: Psychologist*, Hodder & Stoughton, London, 1979.

10. Robert B. Joynson, *The Burt Affair*, Routledge, London, 1989.

11. Rose, Lewontin and Kamin, *Not in Our Genes*, pp. 277 & 285.

12. ibid., p. 116.
13. Thomas J. Bouchard *et al.*, 'Sources of psychological differences: the Minnesota study of twins reared apart', *Science*, vol. 250, no. 4978, 1990, pp. 223–8.
14. Walter Bodmer and R. McKie, *The Book of Man: The Quest to Discover our Genetic Heritage*, Little, Brown & Co, London, 1994, p. 132. Also published by Scribner, New York, as *The Book of Man: The Human Genome Project and the Quest to Discover our Genetic Heritage*.
15. ibid., p. 134.
16. Segal, *Entwined Lives*, is the latest one. Nancy Segal was formerly Assistant Director of the Minnesota Study Center.
17. Bouchard *et al.*, 'Sources of psychological differences', p. 227.
18. ibid.
19. A.R. Jensen, 'The psychometrics of intelligence', in *The Scientific Study of Human Nature*, ed. H. Nyborg, Pergamon, Oxford, 1997, pp. 221–39.
20. Sara J. Shettleworth, 'Memory in food-hoarding birds', *Scientific American*, vol. 248, no. 3, 1983, pp. 86–94.
21. David L. Wodrich, *Children's Psychological Testing: A guide for Nonpsychologists*, 3rd edn, Paul H. Brookes, Baltimore, 1997. An excellent book for non-specialists in general. Devotes several pages to WISC-III. The author is clinical director of child psychology at the Phoenix Children's Hospital in Arizona.
22. Stephen J. Gould, *The Mismeasure of Man*, Penguin, London, 1981.
23. Experimental studies are described by N. Brody, 'Malleability and change in intelligence', in *The Scientific Study of Human Nature*, ed. H. Nyborg, Pergamon, Oxford, 1997, pp. 311–30. The general conclusion is also mentioned in the report of the taskforce on intelligence (note 28).
24. The gradual improvement in test performance, which has been recorded in many technologically advanced countries, is known as the Flynn effect, after the psychologist who first described it. It is discussed briefly in the report of the taskforce on intelligence.
25. This is because scores on intelligence tests are usually converted to a scale on which the mean is 100 and the standard deviation 15. Standard deviation is a measure of the

spread or scatter of scores about the mean. In a normal distribution, about 68 per cent of the area within the bell curve is
enclosed within one standard deviation on each side of the
mean (the central axis of the curve) and 95 per cent within
two standard deviations.

26. R.J. Herrnstein and C. Murray, *The Bell Curve*, Free Press,
New York, 1994.

27. Steven Fraser (ed.), *The Bell Curve Wars: Race, Intelligence and
the Future of America*, Basic Books, New York, 1995.

28. Ulric Neisser *et al.*, 'Intelligence: knowns and unknowns',
report of a taskforce established by the American Psychological Association, *American Psychologist*, vol. 51, no. 2, 1996,
pp. 77–101.

29. 'Exploring intelligence', *Scientific American* (quarterly special
issue), vol. 9, no. 4, Winter 1998. A wide-ranging review with
many up-to-date references.

30. Wodrich, *Children's Psychological Testing*, p. 322.

31. H.J. Eysenck, *Race, Intelligence and Education*, Maurice Temple
Smith, London, 1971, p. 34.

CHAPTER 13   GENES AND BEHAVIOUR

1. Jerome Kagan, *Galen's Prophecy: Temperament and Human
Nature*, Basic Books, New York, 1994, p. 2. The title of this
book derives from Galen's belief that inherited constitution
played a part in producing melancholic and sanguine
individuals.

2. Charles S. Carver and M.F. Scheier, *Perspectives in Personality*,
3rd edn, Allyn & Bacon, Needham Heights, MA, 1996,
pp. 70–1.

3. H.J. Eysenck and S. Rachman, 'Personality dimensions', in
Readings in Personality, eds H.N. Mischel and W. Mischel,
Holt, Rinehart & Winston, New York, 1973, pp. 27–30.

4. C.R. Brand, 'Hans Eysenck's personality dimensions: their
number and nature', in *The Scientific Study of Human Nature*,
ed. H. Nyborg, Pergamon, Oxford, 1997, pp. 17–35.

5. Jerome Kagan, D. Arcus and N. Snidman, 'The idea of
temperament: where do we go from here?', in, *Nature,
Nurture and Psychology*, eds R. Plomin & G.E. McClearn,
American Psychological Association, Washington, 1993,
pp. 197–210.

6. R.M. Stelmack, 'The psychophysics and psychophysiology of extraversion and arousal', in *The Scientific Study of Human Nature*, pp. 388–403.

7. Kagan, Arcus and Snidman, p. 206.

8. Daniel Kimble, *Biological Psychology*, 2nd edn, Harcourt Brace Jovanovich, Fort Worth, TX, 1992, p. 394.

9. John B. Watson, *Behaviourism*, Norton, New York, 1925, p. 82 (p. 104 in the 2nd edition reprinted by Norton in 1970).

10. B.F. Skinner, *Science and Human Behaviour*, Macmillan, New York, 1953, p. 35.

11. ibid., p. 66. Skinner explained that he used the word 'conditioning' because it had already become established in learning theory.

12. ibid., p. 93.

13. B.F. Skinner, *Walden Two*, Macmillan, New York, 1948, with many later printings. The title is a reference to Henry Thoreau's *Walden, or Life in the Woods*, which was published in 1854.

14. Kagan, *Galen's Prophecy*, p. 12.

15. Nikolaas Tinbergen, *The Study of Instinct*, Oxford University Press, Oxford, 1951. Reissued, with new preface, 1989.

16. Irving Kupfermann, 'Genetic determinants of behaviour', in *Principles of Neural Science*, 3rd edn, eds E.R. Kandel, J.H. Schwartz and T.M. Jessel, 1991, Elsevier, New York, pp. 987–96. This massive book is a wide-ranging introductory text and quite accessible to non-specialists.

17. Charles Darwin, *The Origin of Species*, Penguin, London, 1968, p. 234.

18. Pioneering research on the American white-crowned sparrow was carried out by Peter Marler at the University of California in the 1960s and 70s. A brief summary appears in David McFarland, *Animal Behaviour*, Longman, Harlow, 1985, pp. 34–6.

19. The research papers on wagtails and sparrows are referred to in *An Introduction to Behavioural Ecology*, 2nd edn, eds J.R. Krebs and N.B. Davies, Blackwell, Oxford, 1987, pp. 107–8 and 125.

20. B.F. Skinner, *Verbal Behaviour*, Appleton-Century-Crofts, New York, 1953.

21. Noam Chomsky, 'A review of B.F. Skinner's *Verbal Behaviour*', *Language*, vol. 35, no. 1, 1959, pp. 26–58. A reprint of

this paper appears in *The Structure of Language: Readings in the Philosophy of Language*, eds J.A. Fodor and J.J. Katz, Prentice-Hall, Englewood Cliffs, NJ, 1964, pp. 547–78.

22. Steven Pinker, *The Language Instinct*, William Morrow, New York, 1994. Also published by Penguin, London, 1995.

23. Brian Butterworth, *The Mathematical Brain*, Macmillan, London, 1999.

24. 'Why Einstein was Einstein', an 'In brief' report, *Scientific American*, vol. 281, no. 3, p. 20. Einstein's brain was placed in preservative within hours of his death.

25. Numerous research findings are relevant here. Many are listed by John Tooby and Leda Cosmides, University of California, Santa Barbara, 'The psychological foundations of culture', in *The Adapted Mind: Evolutionary Psychology and the Generation of Culture*, eds J. Barkow, L. Cosmides and J. Tooby, Oxford University Press, New York, 1992, pp. 19–136. See especially pp. 71–2.

26. Research papers on the barn swallow by Anders Moller are referred to by M.J. Ryan, 'Sexual selection and mate choice' in *Behavioural Ecology: An Evolutionary Approach*, 4th edn, eds J.R. Krebs and N.B. Davies, Blackwell Science, Oxford, pp. 179–202.

27. The term 'Standard Social Science Model', which is now quite widely used in a critical sense, was coined by Tooby and Cosmides. See note 25.

28. Anthony Giddens, *Sociology*, 3rd edn, Polity Press, Cambridge, in association with Blackwell, Oxford, 1997, p. 22.

CHAPTER 14   THE SOCIAL DIMENSION

1. Richard Dawkins, *The Selfish Gene*, 2nd edn, Oxford University Press, Oxford, 1989, p. 19.

2. Michael T. Ghiselin, *The Economy of Nature and the Evolution of Sex*, University of California Press, Berkeley, 1974.

3. Kim Sterelny and P.E. Griffiths, *Sex and Death: An Introduction to Philosophy of Biology*, University of Chicago Press, Chicago, 1999, p. 318.

4. Exactly how feathers arose from reptilian scales is still unclear, but in the 1960s grafting experiments in embryo birds showed that, during the early stages of development, the same skin cells can produce either scales or feathers according to location on

the body. Donald S. Farner and J.R. King, eds, *Avian Biology*, vol. 2, Academic Press, New York, 1979, pp. 48–9.

5. Steven Rose, R. Lewontin and L. Kamin, *Not in Our Genes*, Penguin, London, 1984, p. 249.

6. Rape in humans and other species has recently become the subject of a book: Randy Thornhill and C. Palmer, *A Natural History of Rape*, MIT Press, Cambridge, MA, 2000.

7. Marshall Sahlins, *The Use and Abuse of Biology: An Anthropological Critique of Sociobiology*, University of Michigan Press, Ann Arbor, 1976.

8. Diane F. Halpern, *Sex Differences in Cognitive Ability*, Lawrence Erlbaum, Hillsdale, NJ, 1992, p. 7.

9. According to Bronislaw Malinowski, the founder of social anthropology, there were people in the Trobriand Islands, even during the twentieth century, who did not consider copulation to be the cause of pregnancy. Cited in Sahlins, *Use and Abuse of Biology*, p. 37.

10. Anne Moir and D. Jessel, *A Mind to Crime: The Controversial Link Between Mind and Criminal Behaviour*, Michael Joseph, London, 1995, p. 14.

11. Elliot Aronson, *The Social Animal*, 8th edn, Worth Publishers, New York, 1999, pp. 404–5. From research reported by R.C. Clark and E. Hatfield in 1989.

12. David M. Buss *et al.*, 'Sex differences in jealousy: evolution, physiology, and psychology', *Psychological Science*, vol. 3, no. 4, 1992, pp. 251–5. The subject is covered more widely in Professor Buss' recent book: *The Dangerous Passion: Why Jealousy is as Necessary as Love and Sex*, Free Press, New York, 2000.

13. Alison Motluk, 'Killing off an archetype', *New Scientist*, vol. 166, no. 2238, 2000, p. 9.

14. *Numbers* 32, 17 & 18.

15. Katherine Ralls, J.D. Ballou and A. Templeton, 'Estimates of lethal equivalents and the cost of inbreeding in mammals', a paper from 1988 reprinted in *Readings from Conservation Biology*, ed. D. Ehrenfeld, Blackwell Science, Oxford 1995, pp. 192–200.

16. Stephen T. Emlen, 'Family dynamics in social vertebrates', in *Behavioural Ecology*, eds J.R. Krebs and N.B. Davies, Blackwell, Oxford, 1997, pp. 228–53.

17. David Barash, *Sociobiology and Behaviour*, 2nd edn, Hodder & Stoughton, London, 1982, p. 257.
18. Although the idea is sound, in practice it is very hard to measure. We cannot know, for example, how many offspring a young helper might have had if he or she had not died saving siblings.
19. For a comprehensive review see Peter G. Hepper, *Kin Recognition*, Cambridge University Press, Cambridge, 1991.
20. The ESS concept is now established as a way of analysing decision-making where the pay-offs depend on the ratios of participants and non-participants. It was introduced by John Maynard Smith in 1972, and elaborated in his book: *Evolution and the Theory of Games*, Cambridge University Press, Cambridge, 1982.
21. Gerald S. Wilkinson, 'Food sharing in vampire bats', *Scientific American*, vol. 262, no. 2, 1990, pp. 64–70.
22. Anne E. Pusey and C. Packer, 'The ecology of relationships', in *Behavioural Ecology*, eds J.R. Krebs and N.B. Davies, pp. 254–83.
23. Daniel J. Povinelli and L.R. Godfrey, 'The chimpanzee's mind: how noble in reason? how absent of ethics?', in *Evolutionary Ethics*, eds M.H. Nitecki and D.V. Nitecki, State University of New York Press, New York, 1993.
24. James S. Chisholm, *Death, Hope and Sex: Steps to an Evolutionary Ecology of Mind and Morality*, Cambridge University Press, Cambridge, 1999, pp. 128–9.
25. ibid.
26. Jerome Barkow, L. Cosmides and J. Tooby, *The Adapted Mind: Evolutionary Psychology and the Generation of Culture*, Oxford University press, New York, 1992, p. 199.
27. Sterelny and Griffiths, *Sex and Death*, p. 345.
28. Henry Plotkin, *The Nature of Knowledge*, Allen Lane, London, 1994, p. 208.
29. Thomas H. Huxley, 'Evolution and ethics'. An extract from the 1893 publication appears in *Evolution*, ed. M. Ridley, Oxford University Press, Oxford, 1997, pp. 395–8.
30. Richard Dawkins, *The Selfish Gene*, 2nd edn, Oxford University Press, New York, 1989, p. 2.
31. Chisholm, *Death, Hope and Sex*.
32. H.J. Muller, 'The penalty for relaxing natural selection', in

Evolution, ed. M. Ridley, Oxford University Press, Oxford, 1997, pp. 341–6.

CHAPTER 15   OLD BRAINS, NEW BRAINS

1. Bryan Kolb and I.Q. Whishaw, *Fundamentals of Human Neuropsychology*, 4th edn, W.H. Freeman, New York, 1996, p. 65. A textbook I used extensively for this chapter.
2. John Parnavelas, 'The human brain: a hundred billion connected cells', in *From Brains to Consciousness?*, ed. S. Rose, Allen Lane, London, 1998, pp. 18–32.
3. Kolb and Whishaw, *Fundamentals*, p. 89.
4. The so-called classical transmitters are built in our bodies from simpler substances in food. They include acetylcholine, dopamine and serotonin.
5. Jon H. Kaas and A. Reiner, 'The neocortex comes together', *Nature*, vol. 399, no. 6735, 1999, pp. 418–19.
6. Kenneth V. Kardong, *Vertebrates: Comparative Anatomy, Function, Evolution*, 2nd edn, WCB/McGraw-Hill, Boston, MA, 1998, pp. 616–30.
7. Eric R. Kandel and T. Jessel, 'Early experience and the fine tuning of synaptic connections', in *Principles of Neural Science*, 3rd edn, eds E.R. Kandel, J.H. Schwartz and T.M. Jessel, 1991, Elsevier, New York, pp. 945–58.
8. Peter Collins, 'The field workers', *New Scientist*, vol. 164, no. 2224, 2000, pp. 36–9.
9. Joseph E. LeDoux, 'Emotion, memory and the brain', *Scientific American*, vol. 270, no. 6, 1994, pp. 32–9.
10. Kolb and Whishaw, *Fundamentals*, pp. 305–33.
11. ibid., p. 362.
12. Eleanor Maguire, *et al.*, 'Navigation-related structural change in the hippocampi of taxi drivers', *Proceedings of the National Academy of Sciences*, vol. 97, no. 8, 2000, pp. 4398–403. For an earlier report on Dr Maguire's work see Nell Boyce, 'The Mall, where's that?', *New Scientist*, vol. 155, no. 2099, 1997, p. 16.
13. Donald O. Hebb, *The Organization of Behaviour*, John Wiley, New York, 1949.
14. Much of the rest of the neocortex (about 55 per cent of it in primates) is involved with vision as well.
15. Semir Zeki, 'The visual image in mind and brain', *Scientific*

American, vol. 267, no. 3, 1992, pp. 42–50. This is a special issue of *Scientific American* mainly devoted to mind and brain.

16. ibid., p. 48.
17. Antonio R. Damasio and H. Damasio, 'Brain and language', *Scientific American*, vol. 267, no. 3, 1992, pp. 63–71.
18. Kolb and Whishaw, *Fundamentals*, p. 205.
19. Angelo Bisazza *et al.*, 'Right pawedness in toads', *Nature*, vol. 379, no. 6564, 1996, p. 408. For a broad treatment of the topic see John Bradshaw and L. Rogers, *The Evolution of Lateral Asymmetries, Language, Tool Use, and Intellect*, Academic Press, San Diego, 1993.
20. Michael S. Gazzaniga, 'The split brain revisited', *Scientific American*, vol. 279, no. 1, 1998, pp. 34–9.
21. ibid.
22. Sally P. Springer and G. Deutsch, *Left Brain, Right Brain*, 5th edn, W.H. Freeman, New York, 1998, p. 39.
23. James W. Kalat, *Biological Psychology*, 6th edn, Brooks/Cole, Pacific Grove, CA, 1998, p. 316.
24. Louis Gooren, Professor of Endocrinology, Free University of Amsterdam, interviewed on *The Health Report*, Radio National, Australia, 4 Sept, 2000.
25. Doreen Kimura, 'Sex differences in the brain', *Scientific American*, vol. 267, no. 3, 1992, pp. 80–7.
26. Kalat, *Biological Psychology*, p. 306.
27. Kolb and Whishaw, *Fundamentals*, p. 229.
28. Anne Moir and D. Jessel, *A Mind to Crime*, Michael Joseph, London, 1995, p. 46.
29. Susan Greenfield, 'How might the brain generate consciousness?', in *From Brains to Consciousness*, ed. S. Rose, Allen Lane, London, 1998, pp. 210–27.
30. Wolf Singer, 'Consciousness from a neurobiological perspective', in Rose, *From Brains to Consciousness*, pp. 228–45.
31. Igor Aleksander, 'A neurocomputational view of consciousness', in Rose, *From Brains to Consciousness*, pp. 180–99.
32. Roger Penrose, *The Large, the Small and the Human Mind*, Cambridge University Press, Cambridge, 1997.
33. David J. Chalmers, *The Conscious Mind*, Oxford University Press, New York, 1996.
34. Mary Midgley, 'One world but a big one', in Rose, *From Brains to Consciousness*, pp. 246–70.

35. Antonio R. Damasio, 'How the brain creates the mind', *Scientific American*, vol. 281, no. 6, 1999, pp. 74–9.

CHAPTER 16   THE FUTURE IS COMING
 1. UNFPA, *The State of The World Population 1999*, UNFPA, New York, 1999, p. 3. An annual report by the UN Population Fund.
 2. ibid, p. 25.
 3. Steve Jones, *The Language of the Genes*, HarperCollins, London, 1993, p. 220.
 4. Arthur P. Mange and E.J. Mange, *Genetics: Human Aspects*, 2nd edn, Sinauer Associates, Sunderland, MA, 1990, p. 448.
 5. Steve Jones, *Almost Like a Whale: The Origin of Species Updated*, Doubleday, London, 1999, p. 339.
 6. Beth Martin and M. Day, 'Fresh alarm over threatened sperm', *New Scientist*, vol. 153, no. 2064, Jan. 1997, p. 5.
 7. Steve Jones, *Almost Like a Whale*, p. 352.
 8. News reports in late 1999 on completion of the United Nations International Decade for Disaster Reduction.
 9. Indra de Soysa and P. Gleditsch, *To Cultivate Peace: Agriculture in a World of Conflict*, Report of the International Peace Research Institute (PRIO), Oslo, Norway, 1999, p. 13.
10. Michael Renner, 'Transforming security', in *State of the World 1997*, eds L. Brown *et al.*, Earthscan Publications, London, 1997, pp. 115–31.
11. Speech by the UN High Commissioner for Refugees at the 46th session of the Executive Committee, Geneva, 16 October 1995.
12. George M. Woodwell, 'The effects of global warming', in *Global Warming: The Greenpeace Report*, ed. J. Leggett, Oxford University Press, Oxford, 1990, p. 128.
13. Jodi L. Jacobson, 'Holding back the sea', in *State of the World 1990*, eds L. Brown *et al.*, Earthscan Publications, London, 1990, pp. 79–97.
14. Woodwell, 'The effects of global warming'.
15. Renner, 'Transforming security'.
16. *The Courier Mail*, Brisbane, 19 April 2000, p. 26.
17. Julian Huxley, *Essays of a Humanist*, Chatto & Windus, London, 1964, p. 92.
18. *The Weekly Telegraph*, London, no. 441, 5 January 2000, p. 21.

The Gallup poll was conducted for this newspaper and the results were analysed by Anthony King, Professor of Government at Essex University.

19. Steve Bruce, *Religion in the Modern World: From Cathedrals to Cults*, Oxford University Press, Oxford, 1996, p. 38–9. A clearly written analysis by the Professor of Sociology at the University of Aberdeen.

20. ibid., pp. 169–95.

21. ibid., p. 210.

22. ibid., p. 225.

23. Part of a formal statement, cited in the *Encyclopaedia Britannica Yearbook*, 1996, p. 306.

24. The decision by the Kansas Board of Education to take evolution out of the state school standards for science was widely reported. For typical responses in science journals see *Science*, vol. 285, no. 5431, 1999, p. 1186; and *Scientific American*, vol. 281, no. 4, 1999, p. 18. A special feature on creationism as a force in society appears in *New Scientist*, vol. 166, no. 2235, 2000, pp. 32–48. See also the editorial of that issue, p. 3.

25. George Bishop, 'The religious worldview and American beliefs about human origins', in *The Public Perspective*, vol. 9, no. 5, 1998, pp. 39–44. *The Public Perspective* is a publication of the Roper Center for Public Research, University of Connecticut, Storrs, CT.

26. Mark Juergensmeyer, *The New Cold War? Religious Nationalism Confronts the Secular State*, University of California Press, Berkeley, 1993, p. 191.

27. Edward J. Larson and L. Witham, in correspondence to *Nature*, vol. 394, no. 6691, 1998, p. 313.

SELECT BIBLIOGRAPHY

Brown, L. *et al. State of the World 1999* (Worldwatch Institute report, special millennium edition), Earthscan, London, 1999

Damasio, A. *The Feeling of What Happens: Body, Emotion and the Making of Consciousness*, Vintage, London, 2000

Dawkins, R. *The Blind Watchmaker*, Longman, London, 1986

Foley, R. *Humans Before Humanity*, Blackwell, Oxford, 1995

Henry, J. *The Scientific Revolution and the Origins of Modern Science*, Macmillan, London, 1997

Jones, S. *The Language of the Genes*, Flamingo, London, 1994

Jones, S., Martin, R. and Pilbeam, D. *The Cambridge Encyclopedia of Human Evolution*, Cambridge University Press, Cambridge, 1992

Margulis, L. and Sagan, D. *What is Life?*, Simon & Schuster, New York, 1995

Moore, J.A. *Science as a Way of Knowing: The Foundations of Modern Biology*, Harvard University Press, Cambridge, MA, 1993

Pinker, S. *The Language Instinct*, William Morrow, New York, 1994. Also published by Penguin, London, 1995

Redfern, R. *Origins: Evolution of Continents, Oceans and Life*, Cassell, London, 2000

Rees, M. *Before the Beginning: Our Universe and Others*, Addison Wesley, Reading, MA, 1997

Ridley, M. *Mendel's Demon: Gene Justice and the Complexity of Life*, Weidenfeld & Nicolson, London, 2000

Roux, Georges, *Ancient Iraq*, 3rd edn, Penguin, London, 1992

Sagan, C. *The Demon-Haunted World: Science as a Candle in the Dark*, Random House, New York, 1996

Segerstråle, U. *Defenders of the Truth: The Battle for Science in the Sociobiology Debate and Beyond*, Oxford University Press, Oxford, 2000

Simmons, I.G. *Changing the Face of the Earth*, 2nd edn, Blackwell, Oxford, 1996

White, M. and Gribbin, J. *Darwin: A Life in Science*, Dutton, New York, 1995

Wilson, E.O. *Consilience: The Unity of Knowledge*, Little, Brown & Co., London, 1998

Index

action potential, *see* nerve impulse
adaptations,
   in evolutionary theory, 73, 77, 78, 82, 85
   examples of, 118, 126, 135, 208
adultery, 208
African Eve, 100, 101
African replacement hypothesis, 100, 101
African Rift, 106–7
Agassiz, Louis, 143
ageing, 153, 246–7
Agenda 21, 166
   early 127, 128, 129–31, 135
   and global diet, 154–5, 156–7; *see also* Green Revolution
Aleksander, Igor, 235
alleles, 32–3, 242
altruism, 85–6, 210, 213; *see also* cooperation, reciprocity
American Civil War, 143
aminoacids, 26–7
amoeba, 51, 74
anarchism, 146–7
animism, 2–3
anthropoids, 93
apes, 93–4, 124, 207; *see also* chimpanzee, gorilla, orang utan
appendix, 238
Aquinas, Saint Thomas, 9
arable land, 159; *see also* grainland
Ardrey, Robert, 151
Aristotle, 6–7, 13
arousal hypothesis, 188
asteroids, 37
atmosphere, 24, 37, 45, 47, 122

atoms, 21–2
Augustine, Saint, 8
Australian Aborigines, 2, 125, 127, 136, 240
australopithecines, 95–6, 97, 98, 108
bautism, 214
Avicenna, 8

baboons, 108, 113, 117
Bacon, Sir Francis, 10–11
bacteria,
   and DNA, 31, 45–6,
   and nutrients, 50
   and organelles, 51
   and oxygen, 47
   size of, 46
   as survivors, 18–19, 38–9, 40; *see also* chloroplasts, cilia, mitochondria, stromatolites
balance of nature, 126, 127
Beagle, HMS, 58
Beecher, Henry Ward, 143
bee-eater, 86
behaviourism, 191–3,
big bang, 34–5
bipedalism, 94, 108
birds, 70, 115, 164, 174, 197, 200, 203, 209; *see also* bee-eater, Clark's nutcracker, feathers, imprinting, shearwaters, sparrows, wagtails
birdsong, 196–7
birth control, 154, 205; *see also* contraception
birth weight, human, 70

blood, pigments, 13
Boas, Franz, 146
bonobo, 111
brain,
    chemicals, 219
    and consciousness, 234–6
    development and structure,
        220–2
    and energy needs, 109–10
    functions of parts, 223–4
    and language, 229, 231
    learning and memory, 224–7
        passim
    left and right, 229–31
    male and female, 231–4
    split brain studies, 230–1
Bruce, Steve, 248, 249
Bruno, Giordano, 9
burial practices, 130–1, 152
Burt, Cyril, 175, 176

Cambrian, 54, 90
cannibalism, 114
carbon, 24
carrying capacity, 125, 127, 128
catastrophism, 59; see also
    uniformitarianism
Catholic, 8, 9, 11, 143, 154, 248
cave painting, see Stone Age art
Cech, Tom, 41
cells, 20, 44–55 passim
    division of, 29, 153, 170, 174
    sex (gametes), 29
Cenozoic, 106
cerebellum, 220, 221, 223
Chalmers, David, 236
Chambers, Robert, 57
chance in evolution, 69
chimpanzee, 94, 111, 115–16, 117,
    136, 213–14; see also bonobo
Chinese,
    inventions, 12
    agriculture, 129, 135
    infants, 187
chloroplasts, 48, 51

Chomsky, Noam, 198
chromosomes, 25–6, 28–32 passim
    sex, 232, 233
Church, 6, 9, 59, 139, 140, 141,
    250
cilia, 52
cistron, 31, 84
city-states, 132–3
Clark's nutcracker, 180, 227
classification, see taxonomy
coal, 12, 137
COBE satellite, 35
Collard, Mark, 97–8
competitive exclusion, 103
conditioning,
    classical, 190, 223
    operant, 192, 193
    fear, 224; see also emotion
consciousness, see brain
continental drift, 89–90
contraception, 241, 248
cooperation, 211
Copernicus, 9
copper tools, 132
corpus callosum, 229, 230
correlation, 178
Cretaceous, 91
Crick, Francis, 26, 27
critical period, 223, 233; see also
    sensitive periods
Cro-Magnons, 119
cultural universals, 150
cuneiform script, 133
Cuvier, Georges, 59
cystic fibrosis, 32

Damasio, Antonio, 236
Dart, Raymond, 95
Darwin, Charles, 57–68 passim,
    141, 147
    remarks and observations by,
        67, 81, 87, 141, 142, 147,
        158, 196; see also Darwinism
Darwin, Erasmus, 56
Darwinism, 69

social, 144, 204
Soviet creative, 148
Davies, Paul, 34
Dawkins, Richard, 77–8, 80, 84, 100, 176, 216, 251
DDT, 161
De Duve, Christian, 48
death, programmed, 54
Dennet, Daniel, 82, 83
Descartes, René, 11, 138, 193, 234
diet, 126, 154, 156, 157
disease, 16, 136, 242
dive response, 195–6
DNA, 25–33 *passim*, 38, 42, 168–70
    mitochondrial, 100, 104; *see also* telomeres
dogs, 11, 61, 65, 73–4, 130
domestic animals, 129–30, 136
Driesch, Hans, 19
dualist views, 6, 11, 138, 234
Dubois, Eugene, 98

Early Dynastic, 133
Earth, 36, 90, 106, 122
ecological separation, 107, 158; *see also* competitive exclusion, speciation
economics, 167, 203–4, 246
Egypt, ancient, 3–4
Ehrlich, Paul, 166
Einstein's brain, 199
Eldredge, Niles, 81
electrons, 21–2, 23, 24
elephant seal, 206
emotions, 215–16, 224
encephalisation quotient, 109–10, 115
energy pyramid, *see* food chains
energy, 22
    and brain requirement, 109–10
    dietary, 113, 156
    and human needs, 127, 134, 137
    and lifestyle, 159
    and natural selection, 197

from the Sun, 122, 158; *see also* metabolism, food chains, photosynthesis
English Civil War, 138
environmental degradation, 162
environmental determinism, 176, 193
enzymes, 24–5
ESS, 212
ethics, 216–17
ethnic mixing, *see* intermarriage
ethology, 150
eugenics, 145–6
eukaryotes, 48, 49, 50
evolution,
    early concepts of, 56–7
    modern synthesis, 68–9, 83; *see also* disease, natural selection
evolutionary psychology, 217; *see also* sociobiology
extinction, 107, 164, 165
eyes, evolution of, 74–6; *see also* vision
Eysenck, H.J., 184, 187, 188

false memory, 230–1
family size, 241, 242; *see also* fertility rate
farming, *see* agriculture
feathers, 87, 203
feminists, 112–13, 203
fertile crescent, 129, 132
fertility rate, 165; *see also* family size, population
fire, 120–1, 125
fishing, 157
fitness, 210
fixed action pattern, 195
flatworms, 54
Foley, Robert, 110
food chains, 123, 157, 161
Franklin, Rosalind, 26
Freud, Sigmund, 224
frontal lobes, 221, 224, 225–6
Furedi, Frank, 167

Gage, Phineus, 225
Galen, 7–8, 186
Galileo, 9
Galton, Francis, 145, 174, 175, 179
gametes, *see* cells
ganglia, 220
genes, 31–3, 75, 76, 77, 166, 168–74 *passim*, 245
  and behaviour, 185–8; *see also* instinct, nature–nuture relationships, personality
  regional distribution of, 238–41 *passim*
  regulatory, 170
  selfish, 80; *see also* cistron, replicator, units of selection
genetic determinism, 79, 84, 193
genetic diversity, 165, 166, 241
genome,
  of nematode, 32; *see also* Human Genome Project
genotype, 67, 172, 179, 207, 208
gene pool, human, 239, 243
Ghiselin, Michael, 202
giand panda, 105
Gilbert, William, 11
glaciation, 106
global warming, 243–4, 245
gorilla, 113, 117
grainland, 156
Gould, Stephen Jay, 80–4 *passim*, 101
Greeks, ancient, 4–7 *passim*, 56
Green Revolution, 155, 156, 161
Greenfield, Susan, 234
group selection, 85, 86, 87, 212

Hamilton, William, 86, 210
handicap principle, 200
Harvey, William, 11, 13
Hawking, Stephen, 34
Hebb, Donald, 227
Henslow, John, 58
heritability, 173–4

hippocampus, 221, 226, 227, 229, 234
Hippocrates, 5, 186
HIV, 136, 242
Hobbes, Thomas, 138–9
hominids, 94–8 *passim*
Homo,
  early ecology of, 109–20 *passim*, 124–6; *see also* Cro-Magnon, *Homo* species, Java Man, Neanderthals, Peking Man
  as a genus, 97–8
Homo species,
  Homo erectus, 99, 101, 103, 109
  Homo ergaster, 99
  Homo habilis, 97, 98, 112
Hooke, Robert, 20
Hooker, Joseph, 62, 63
hormones, 219, 221, 232, 233
Hoyle, Fred, 34, 40
Human Genome Organization, 169
Human Genome Project, 169–71, 242
human nature, 138, 139, 140, 149
Hume, David, 140, 141, 216
humours, doctrine of, 5, 7
hunting and gathering, 124–8 *passim*, 131, 135
Hutton, James, 59
Huxley, Julian, 68, 154
Huxley, T.H., 142
hypothalamus, 221, 224
hypotheses, 16

Ibn Sina, *see* Avicenna
ice ages, *see* glaciation
ignorance, categories of, 64, 67
immune system, 211
imprinting, 196
incest, 209
inclusive fitness, 210

industrial revolution, 12
infant mortality, 127, 152
instinct, 193–9 *passim*
intelligence, 179–84 *passim*, 216
    testing, 175, 177, 179, 180–3
intermarriage, 240
ions, 23, 45
IQ, *see* intelligence
iron smelting, 134
irrigation, 132, 163

James, William, 235
Japanese macaque, 111
Java Man, 98, 99
jaws, 91
jellyfish, 53
Jeon, Kwang, 51
Jericho, 131
Jones, Steve, 31, 146
Jung, Carl, 187
Jurassic, 91, 92
Just So Stories, 82, 200, 202

Kalahari bushmen, 125, 127
Kamin, Leon, 175, 176
Kauffman, Stuart, 43
Kelvin, Lord, 67
kibbutzim, 209
kinship, 86, 149; *see also* nepotism
Kohler, Wolfgang, 194
Kropotkin, Peter, 146–7
Kruschchev, Premier, 149

Lamarck, Jean Baptiste de, 56
Lamarckism, 56–7, 68, 148
language, 97, 119, 121, 133,
    198–9; *see also* brain
Leakey, Louis, 97
leapfrogging, 162
learning, *see* brain, conditioning,
    instinct, trial and error learning
Lewontin, Richard, 81–2, 176
life,
    expectancy, 138, 152
    origins of, 37–43 *passim*

problems of defining, 18–21
    passim
limbic system, 224, 226
Linnaeus, Carolus, 92
lipids, 24, 44
    and membranes, 50
liver cells, 21
Locke, John, 140, 192
Lorenz, Konrad, 196
Lucy, 96
Lyell, Charles, 59, 60, 62, 63, 65
Lysenko, Trofim, 147–8

Malthus, Thomas, 62, 157
mammals, origin, 91–2
Margulis, Lynn, 51, 52, 54
Marx, Karl, 147, 176
memory, 225–7, 229; *see also*
    Clark's nutcracker, false
    memory
Mendel, Gregor, 66–7
menstrual cycle, 118, 207
Mesolithic, 128–9
Mesopotamia, 132, 133, 134, 135,
    163
metabolism, 22, 123
Michurin, Ivan, 148
microliths, 129
Midgley, Mary, 236
Miescher, Friedrich, 25
migration, human, 239–40; *see also*
    refugees
milk digestion, 135–6
Miller, Stanley, 37, 38
Miocene, 94
missing link, 94, 98
MISTRA, 177–8
mitochondria, 48, 51, 99–100
molecular clock 100
molecules, 22–3; *see also* lipids,
    nucleic acids, polymers
morals, 213–16 *passim*; *see also*
    ethics
Morris, Desmond, 151
Murchison meteorite, 40, 44

mutations, 33, 42, 75, 76, 102
    rate of, 75, 102
muttonbirds, *see* shearwaters

national parks, 164
nation-state, first, 133
Natufians, 129, 131
natural selection, 42, 60–2, 64–5,
    69–71, 77, 78, 238, 241
    and behaviour, 198, 205, 207
    and moral values, 213–16
        passim; *see also* adaptation,
        altruism, group selection,
        punctuated equilibria,
        selection pressure, sexual
        selection, units of selection,
Natural Theology, 141
naturalistic fallacy, 216
nature–nurture relationship,
    171–3
Neanderthals, 103–4, 152
neocortex, 222, 223, 226
neo-Darwinism, 69
Neolithic, 127–31 *passim*, 134, 157
nepotism, 210
nerve cell, *see* neuron
nerve impulse, 45, 218, 219
neural tube, 220, 221
neuron, 218, 222–3, 234
neurotransmitter, 219
New Age, 249–50
Nightingale, Florence, 16
'noble savage', 140
normal distribution, 182
nucleic acids, 25–6; *see also* DNA,
    RNA
nucleotides, 26, 29, 30, 38, 42

orang utan, 111
organ systems, 53–4
organelles, 48, 50; *see also*
    chloroplasts, cilia, mitochondria
oxygen, 13, 47

Paley, William, 141

Pangea, 90–1
pangenesis, 65
Panglossian paradigm, 81
parental investment, 207
Pavlov, Ivan, 190
peacock, 87, 115
Peking Man, 98, 99
Penrose, Roger, 236
personality, 185–8
    and temperament, circle of, 186
PET scans, 228
phenotype, 67, 172
pheromones, 118
photosynthesis, 47, 123, 156, 159
Pinker, Steven, 199
pituitary gland, 221
plasma membrane, *see* lipids
plate tectonics, 90, 106
Plato, 5, 6
Pleistocene, 126
pollution, 160–2, 164
polygyny, 117, 206
polymers, 24
Pope, the, 249, 250
population,
    of Britain, 137, 165
    densities, 125, 127, 135, 160
    and diet, 156, 157
    doubling, 153
    and environment, 160, 164, 166,
        244
    global increase, 153–4, 157,
        238–9
    lobby group in Australia, 246
    in Pleistocene, 127
    structure, 154, 247; *see also*
        family size, fertility rate
Portugal, 240
pottery, 129, 131, 132
primates, 93
Prisoners' Dilemma, 212
prosimians, 93
protein, 26–7; *see also* amino acid,
    fishing
punctuated equilibria, 81

rape, 203, 205
reciprocity, 213
reduction division, 29
reductionism, 14, 79, 84
Rees, Martin, 14, 34, 36
refugees, 244–5
regional continuity hypothesis, 101
reinforcement, 192, 193
religion, 2, 8, 10, 131, 132, 134,
    247–51
  in America, 250
  and American scientists, 251
  in Great Britain, 248
  and Hume, 140
  and Spencer, 144; see also
    animism, Catholic, Church,
    Natural Theology, New Age,
    theology
replicators, 84
reproductive isolation, 74–4
ribosome, 30, 42
Ridley, Mark, 84, 88
RNA, 26, 30, 38, 41–3, 170
Rose, Steven, 176
Rousseau, Jean-Jacques, 139–40
Royal Society, 60, 181
Rwanda, 244

Sahlins, Marshall, 204
salinisation, 134, 162–3
Sanger, Margaret, 154
scavenging, 114
Schwaner, Terry, 72–3
science, definition, 12
scientific revolution, 10, 11
sea bed, 90
sea level, 106, 245
selection pressure, 75, 108, 243
self awareness, 115–16
sensitive periods, 196, 199, 233
settlements, early, 130–3
sex,
  reason for, 87–8
  change, 233
  sex differences, see brain

sexual dimorphism, 117
sexual selection, 87
shearwaters, 72
sickle cell disorder, 76
sign stimulus, 195
significance level, 15
Simon, Julian, 164, 166
Singer, Wolf, 235
Skinner, B.F., 192–3, 198
smiling response, 195
social awareness, 115
social evolution, 237
sociobiology, 116, 205, 213, 217;
    see also incest
soul, 2, 6, 9, 11, 138, 194
spandrels, 82
sparrows, 197
Spearman, Charles, 179
speciation, 73–4, 102
Spencer, Herbert, 142, 143, 144
spirochaetes, 52
splitters and lumpers, 99
sponges, 53
SSSM, 201
Srinivasa, Ramanujan, 181
stickleback, 195
Stone Age, 119–20, 121, 128
stone tools, 97, 111, 112, 128–9
Stringer, Chris, 101
stromatolites, 49
subspecies, 93, 102
Sumerians, 133–4
Sumner, William Graham, 144
sun, 36, 122, 158

Taiwan, 209
tapeworm, 136, 160
Tattersall, Ian, 111–12
Taung skull, 95
taxi drivers, 227
taxonomy, 92–3
teleology, 65
telomeres, 153
temperament, see personality
temples, 132–3

temporal lobes, 226
Tertiary, 93
Thales, 4
The Bell Curve, 183–4
The Origin of Species, 61–6 *passim*,
    68, 81, 141, 142, 143, 147, 148,
    194
theology, 10
theory, 16–17
theory of mind, 214; *see also*
    self awareness
therapsids, 91–2
Thorndike, Edward, 191
Thorne, Alan, 101
tiger snakes, 72–3
time-scales, 200–1
Tinbergen, Niko, 195, 200
TMS, 223
towns, 131, 132, 137
tree clearance, 162
trial and error learning, 191
Tudge, Colin, 110
twins, 136, 173–9 *passim*
Tylor, Edward B., 1

Ubaid period, 132
uniformitarianism, 59
units of selection, 80, 83–4
urban environment, 243
urbanisation, 137, 159

vampire bats, 212–13
Vavilov, 148
vervet monkey, 111
Vesalius, Andreas, 9
viruses, 42
vision, 228–9

wagtails, 198
Wallace, Alfred Russell, 63
Wason, Peter, 214
water, 4, 23–4, 163
Watson, James, 26
Watson, John B., 191–2
Wegener, Alfred, 89
Weismann, August, 68
wheels, 132
Wilberforce, Bishop, 142
wildlife conservation, 164–5
Wilkins, Maurice, 26
Wilson, Alan, 100
Wilson, Edward O., 150–1, 176,
    202
Woese, Carl, 39, 43
Wolpoff, Milford, 101–2
Wood, Bernard, 97–8
writing, early, 133
Wundt, Wilhelm, 187
Wynne-Edwards, V.C., 85

Yanomami Indians, 152